Get Into Radio

KU-331-220

Robert C. Elsenpeter

Fender Publishing Company
Seattle Washington

ISBN: 0-9640354-9-9   SAN: 298-1750

Library of Congress Catalog Card Number: 98-22432

Elsenpeter, Robert C., 1970-
       Get into radio / Robert C. Elsenpeter. -- 1st ed.
              p.   cm.
       Includes bibliographical references and index.
       ISBN 0-6940354-9-9 (paperback)
       1. Radio broadcasting--Vocational guidance.   I. Title.
PN1991.E48   1998
384.54'023--dc21   98-22432
                                                             CIP

Books from Fender Publishing Company are distributed by BookWorld Services, Inc. To order additional copies, please call 800-444-2524.

Index by Michelle B. Graye.

Printed in the United States of America

This book is dedicated to Janet, Marilyn and Madeline Elsenpeter; Rowena Pope; Paula Kringle; Deborah Bendix and Amber.

Every good thing in my life is because of these women.

Contents

Note To The Reader

The advice, recommendations, jokes, and other assorted proclamations in this book represent the personal opinions of the writer and, by extension, the publisher. However, neither the writer nor the publisher is responsible for any reactions, events, or other job-seeking adventures that might be inspired by this book. The purpose of this book is to entertain and provide information. This book is not subject to any outside governmental or legal review process. Neither is it a manipulative tool of the totalitarian Establishment, the hungry hounds of political correctness, or America's drab and sappy corporate media machine.

If you have opinions that differ from the ones in this book or you would simply like to correspond with a cool publisher like us, we would like to hear from you. Write to Fender Publishing Company, 1111 East Madison, Suite 460, Seattle, WA 98122.

Don't Go to Law School....

Get Into Radio

Introduction

WGIR starts its broadcast day....

Who hasn't listened to the radio and thought, "Gee, I wish that was me. I know I could do that!" Who hasn't dreamed of reciting witty, fresh, insightful commentary? Perhaps thousands—or even millions—of listeners would be compelled by your voice. Who hasn't lain awake at three a.m., listening to smooth, mellow jazz tunes and heard the deep, soft voice of a disc jockey who calls himself Captain Midnight?

I know I have, and I'll bet you have, too.

Radio careers seem unattainable. They seem so hard to get that many people who probably should be on the radio are not. The unfortunate reality is that many people look at high-profile radio jobs and think they just can't do it. As a result, people who dream of being a disc jockey, a radio talk show host, a news announcer, or any other kind of professional in radio just don't bother to follow their dream. Instead, too many smart and talented young people end up taking the easy way out and going to law school instead, or to medical school.

You want to make money, to do something with your mind, and maybe even to interact with people. But you can do all of that in radio! Of course, it takes a while for the big bucks to start rolling in, but those are the years that you'd be sitting in some boring class, listening to some blowhard tell you what a superior person you are. You know, he tells everyone that. Meanwhile, your debt is piling up and, eventually, you either have to start slicing people up or ruining their lives with litigation in order to make any money at all. Radio people don't have to go down that slimy path. People who work in radio provide the world with entertainment, and they have a great time doing it.

Sure, doctors and—to a far lesser extent—lawyers have accomplished a few good things, maybe. Old ladies run to the doctor when their big toes hurt. The doctor prescribes some disgustingly over-priced drug and, eventually, they feel better. But what have doctors done for you lately?

Lawyers, amazingly, have accomplished a few things, in spite of themselves. They helped create some important civil liberties, and they defend the coolest serial killers. Usually, however, they're busy suing people and companies over stupid things so that they can send out their bills. How about divorced people? Man, they really get a gut-full of "the law." It's so bad that even the lawyers themselves claim to hate one another.

Sometimes students and people who want to change their careers chose law and medical school because they don't know how to get into a fun job. Each year, thousands of young, impressionable minds are irrevocably warped and ultimately washed away from decent society when they set out to become lawyers or doctors. Already, I am mortified to report, there are 656,000 lawyers (one for every 396 people) and 539,000 doctors (one for every 482) in the United States. Comparatively, the Feds tell me there were 50,000 broadcasting jobs.

According to the US Bureau of Labor Statistics, in 1993, 40,302 fresh-faced lawyers came slithering out of the law schools. Also in 1993, 15,531 doctors came crawling out of the medical schools, in search of their first sports car. Those numbers show no sign of decreasing. Do you really want to be among them?

Get those priorities straight

One of the most pathetic things about lawyers and doctors is that they think they are better than the rest of us because they stuck out three or four extra years at some post-graduate school. They didn't learn how to be better people; all they figured out was how slime runs through people's bowels and how to throw old people into the snow when they can't pay their rent.

These turds consider us the ignorant unwashed masses who—thanks to liability and medical insurance—can keep them living the Mercedes-lifestyle with a generous helping of T-bone steaks heaped on top. All their fancy jargon and all their lobbying power in Washington can't convince me that they have anything going on above and beyond pure greed. Allow me to go on a little more, I'm having fun here.

They aren't so smart

Doctors and lawyers would love for you to think they are the most brilliant people on the face of the planet. The following dumb things were actually said by real lawyers while they argued various cases in courts of law.

"Now, doctor, isn't it true that when a person dies in his sleep, in most cases he just passes quietly away and doesn't know anything about it until the next morning?"

Idiot Lawyer: "What happened then?"

Witness: "He told me, he says, 'I have to kill you because you can identify me.'"
Idiot Lawyer: "Did he kill you?"

"The youngest son, the 20-year-old, how old is he?"

Meathead Lawyer: "What is the meaning of sperm being present?"
Coroner: "It indicates intercourse."
Meathead Lawyer: "Male sperm?"
Coroner: "That is the only kind I know."

Dumb-ass Lawyer: "I show you Exhibit 3 and ask you if you recognize that picture."
Witness: "That's me."
Dumb-ass Lawyer: "Were you present when that picture was taken?"

Retarded Lawyer: "Doctor, how many autopsies have you performed on dead people?"
Medical Examiner: "All my autopsies are performed on dead people."

As you can see, even though these jerks had to attend seven years of school, pass the bar examination and, allegedly, spend hours researching legal subtleties and nuances, they're idiots. Plus, their wives won't even touch them, anymore.

Regrettably, many young people decide to attend law school or medical school simply because it's easy. It's easy to sit through a few years of law school, then work in the trenches of some law firm getting practical experience learning how to screw people against the wall.

Ditto doctors.

You could spend thousands of dollars at medical school poking cadavers in the ass and feeling your way through their guts, then you get to go spend time in residency at a hospital getting experience learning how different wounds bleed or develop pus. After that, you'll end up in managed care, learning how to misdiagnose people and overcharge their insurance companies.

You want something better, don't you? You want a fun job, an interesting, non-toxic job that can keep you going. Believe me, when your work is all entertainment, and making sure that people are having a good time, you can damn well expect to find some fun there for you while you're doing it. Why do you think so many lawyers hate their jobs? Why? Because they're boring and evil.

So don't be a parasite of human misery! Do something fun. Get yourself a dream job that will take you places, pay your bills, and make you eager to get up in the morning (or the middle of the night) to get to work.

I'm going to help you get that dream job. Radio can be the greatest rush in the world. Not only will you feel thousands of ears hanging on your every word, sometimes you can even make a difference in peoples' lives. And, unlike doctors, you don't have to look inside people's ears....*yeeech!*

You won't make a fortune right off the bat. It takes time to build up a high salary in radio, but it does happen, especially in the larger markets. But you have to start at the beginning. Even after you've paid your dues, you still won't make what the President's lawyer gets in a day. It's a trade-off to be sure, almost like a pact with the devil. Do you want to sell your soul for instant cash, or do you want adequate money and real job satisfaction? That's a decision you must make for yourself.

Before we get too far into the book, let me tell you we're not just talking about being a disc jockey or a talk-show host. There are lots of other great radio jobs that you can get if you want them. After reading this book, you might decide that, rather than become an on-air personality, you'd like to be a reporter or shoot for production. Maybe you're the type who can sell a ketchup sandwich to a woman in a white dress. In that case, sales is your game.

Down-right fun

Of course, like every other thing in life, a radio career has many "glass half empty/glass half full" aspects to it. Some might find the disc jockey's tendency to move around from one town to another to be a pain in the ass. Others think it is a great way to further your career to see other parts of the world.

The truth is, on-air life is just down-right fun. What can I say? The people who do it are there because they enjoy it. It's not just a way to make ends meet. Radio professionals enjoy knowing that others listen and depend on them for information. Being on the air *is* a big deal. Thousands of people rely on you to deliver their news, weather and sports on the top and bottom of the hour. They rely on you to play the music they want to hear. They rely on you to make them laugh on the way to work and unwind on the way home.

Probably most important, they rely on you to get their girlfriends in the mood, so they can score on Friday night.

There are also fun, exciting jobs that are not on the air.
- Program directors decide how often the public will hear "MacArthur Park" by Donna Summer.

8

- Producers contact and line up compelling guests for the daily talk shows.
- Production engineers put together those funny morning-drive bits.

Though you won't make a big, steaming pile of cash at first, if you do well and maintain a standard of quality and creativity, the ladder of success is yours to climb. Though the heavy hitters like Howard Stern and Rush Limbaugh are few and far between, you *can* someday achieve that status.

I want you to land that radio job. I want you to realize that dream of sitting behind the mike or interviewing a superstar. This book is going to walk you through the process in a very realistic way.

I am not going to blow smoke up your ass. Other "get a job in radio" books are written by broadcasting professors who haven't done a lick of work in the real world for 20 years. They throw tons of vague information at you about theory and the history of plastic radio dials, charge you 20 bucks and make you sort it all out. I'm not going to do that to you.

I will tell you, precisely, what steps you need to take to land that first job. I'll give you the low-down on college, I'll tell you all about getting your voice in shape, and I'll show you how to make a demo tape. From there, I will help you develop your career and climb up to the higher rungs on the ladder to success.

This book maybe a smidge more caustic (by that I mean I swear a lot) than other career books, but that's for a great reason: You're not an idiot. I'm not going to talk to you like you're an idiot, a sixth-grader, or a lawyer.

Other books approach you like you're a mouth-breathing, knuckle-dragging doof who couldn't find his ass with both hands, a flashlight, and a truffle-sniffing pig. I'll give you a little more credit. I won't sugar-coat or paint anything over. I won't hide anything from you just because it might make the radio industry seem bad. I'll arm you with the information you need to know, so you can go in with both barrels loaded.

Some radio folks don't want you to get a radio job because competition is fierce, as it is in any celebrity-oriented industry. For every new Turk who comes down the chute, finding a new job gets that much tougher, so they probably won't share their wealth of knowledge with you. However, I was able to scrounge up some very generous radio professionals who have generously contributed their hard-won wisdom to this book.

I'll tell you what techniques and strategies will work to get you that first radio job, but I'll be real. If you're not cut out for a job on the air, I'll help you figure that out before you spend a dime on schooling. Then, maybe I can steer you toward something you might like better.

Don't take this to mean I'm trying to scare anyone away from following his or her dream. Am I saying you can't get a job in radio at all? No way, not at all. Radio is a big machine with dozens of moving parts. Each one just as valuable and vital as another. If you love the medium and are dying to work in it, then there's a spot for you.

I want you to know you can have *any* job you want, but you have to try. Just buying this book is not enough. It will be hard work. Consider this book a blueprint for achieving your career goal. Consider this book a starting point. You are on the first step of a great journey.

Will the journey be challenging? Yes, it will. But, if you go into this the right way and take as much out of the experience as you can, it will also be very rewarding. Will there be setbacks in your career quest? You bet there will.

But, I'll be giving you lots of tips and tricks to help you get over the big setbacks and steer clear of the little land mines that sneak in here and there.

So get ready and settle in. And whatever you do, don't go to law school! We're going to get you a job in radio.

1. The Careers

The radio biz

Go over to your stereo right now and tune into a station on the dial. Any station will do.

Go ahead. I'll wait.

(Please hum the theme from *Jeopardy* until you have accomplished this task.)

Now, no matter what you're listening to—whether it's news, weather, music, sports, ads, some public service announcement, some Mexican folk song on one of those Spanish stations, *whatever*—it's important to realize that radio is not a one-man operation. You might hear the disc jockeys or announcers all the time, but in reality there are all kinds of staff behind the on-air personalities who set up the pins so the jock can knock them over.

More jobs than you can shake a microphone at

Disc jockeys often gripe about stupid listeners who think it's always request time.

"They don't understand," says Lisa Wright, a St. Paul, Minnesota, disc jockey. "I don't get to pick what I play. It's not up to me. I wish it was, but it isn't. I have to play what the program director tells me to play."

Program directors decide what the disc jockeys play. Further up the radio food chain, upper management listens to its consultants who perform expensive market research.

They then tell the radio station which songs its listeners want to hear.

Hey, I don't like it either. I feel like a dick for calling in now and then, asking my favorite radio station to play "Sex Machine" by James Brown. But if it's not on the play list, it's not getting any air. And it's not only the disc jockeys being jerks either. If they deviate from the play list, it's their ass.

How about those ads that come on the air and annoy the ever-living shit out of us? If I had my way, it would just be music all the time. We'd go from one great song to the next, without hearing about Massengill this or Gold Bond Medicated Powder that. In fact, I'd rather set my hair on fire and put it out with bacon grease than listen to one more dumb-ass commercial.

Unfortunately, radio stations do have to pay their bills. They have to pay for silly things like transmitter towers, staff salaries, and paper clips. To accomplish this, they sell advertising (duh) to anyone who will throw money at the station.

Big business

Radio revenue in 1996 was $12.4 billion and is expected to grow to $23.6 billion by 2006. That's a 6.6 percent increase, according to Kagan Media News, a 28-year-old media research firm. To put all media in comparison, the newspaper industry led the pack ($52.7 billion in '96), but its growth rate is only expected to be 5.6 percent over the next 10 years. The Walt Disney Company is the largest pure media company, with a market capitalization of over $50 billion.

In the July 7, 1997, issue of *Broadcasting and Cable* magazine, Radio Advertising Bureau President Gary Fries gives a good, hearty public-relationsesque explanation for this expected growth, saying it is due to "the ever-increasing number of advertisers who are discovering radio's power to deliver exceptional results."

Though heartily spanking the radio industry's monkey, Fries does make a good point about the present state of the trade. Namely, the industry is doing well. That the radio biz is seeing record profits only means good things for you, the job seeker. Stations want to keep selling, advertisers want to keep buying, and the stations that stay in business might end up giving you a job.

Why does radio advertising rake in so much bread? According to industry estimates, more than 210 million Americans (that's 96 percent of the population) listen to the radio every day. It's a pervasive business. Radio reaches a lot of people; it makes a lot of money; and (here's where it really affects you) the companies hire a lot of people.

A job for everyone

There are all kinds of jobs at a radio station, besides the disc jockey position. Each of them is just as vital (though admittedly not as glamorous) as the jock. They all rely on the others for their existence.

Without sales staff, there would be no money to run the station. Without the production department, there would be no Lexus ads to bring in that money. Without the jocks, there would be no reason to listen to the Lexus ads sandwiched in between the two most popular songs in America today.

I mean, Christ, who wants to produce ads for the Lincoln dealership down the road? I sure as hell don't. I don't even like Lincolns. Old people drive them really slow in the left lane, leaving the turn signal blinking for eight miles after their turn. When used car dealers get around to selling those big-ass boats, you can never get the old-person smell out.

Back to my point.

I might not want to sell ads, I might not want to produce those ads, but someone out there does. Someone wants the big, fat commission check that she'll get from the Lincoln dealership. Some smart aggressive salesperson wants those ads produced. She wants those Lincoln dealerships to become so profitable that they keep on throwing money at the radio station.

Carve yourself a slice

Given the numbers of radio stations and the revenues pulled in each year by the industry, you better believe there is a slice of that pie out there for you. You can have some of it, if you ask for it. In 1996, there were 18,136 radio stations broadcasting in the United States and Canada. I am sorry to report, however, that country music was the most prominent and pervasive format, accounting for 2,948 of the stations.

Though your dream is to get a radio job, do not go into this thinking you will immediately change the world. Don't go into this thinking you will have *carte blanche* to do whatever you want on the air, because that just won't happen. Remember what I said earlier in this chapter. Radio revenues brought in $12.4 billion in 1996. When individual radio stations start thinking about their piece of that giant pie, they want to stick with their consultants' research (for which they pay thousands of dollars).

So, I am sorry to say, as a disc jockey you might end up as some poor little cracker, playing a Snoop Doggy Dogg marathon at three o'clock in the morning in the armpit of Detroit because Mile High Malt Liquor bought the time on your station.

Industry breakdown

A radio station and its jobs can be divided into five general departments: engineering, programming, news, sales, and administration. This is a very general breakdown. No two radio stations will be staffed exactly the same way.

In fact, in small radio stations, the disc jockey might double as the news director and the sales representative. In larger big-wig markets where everyone earns the good dough, there can be dozens of people holding down the fort.

Before we get too deep into the radio careers out there, let me fit you with a bulletproof vest of sorts. The last thing I want to do is scare anyone away from a career in radio. It's what you want to do. But, in order to get your dream job, you will have to reckon with two somewhat unpleasant things:

- Bottom rung of a long ladder, and
- Paying your dues.

The unfortunate fact is that nine out of ten new hires will have to do some menial, bullshit work before they get the good job in radio. It's such a competitive business that starting at the bottom is the best way to get your foot in the door and your name on a check.

Let's take a look at each of the departments and consider the available jobs.

On the Air

The on-air world covers everyone from sportscasters to news people to disc jockeys to talk show hosts. The jobs are, fundamentally, the same, but each has its own twists and turns. These are the jobs I'll bet 99 percent of you picked up this book to find out about. I bet you've heard those guys on the air and said to yourself, "Why not me?"

You are absolutely right. Why not you?

The people who already have these jobs aren't somehow better than you. They don't possess any supernatural talent you don't have (or can't learn). They just don't want you after their jobs. In 1994, there were 50,000 announcing jobs on stations across America. Getting an on-air job isn't easy. Each year thousands of people try to achieve their dream and get their voices heard on the air.

The job is very appealing for many reasons. Some radio hopefuls long for the chance to be the smooth voice coming out of your car stereo. Some want to use their microphone as their personal political pulpit—like Rush Limbaugh who commands 20 million loyal listeners. Some have pseudo-noble, idealistic goals, or so they think, wanting to "make a difference" in their listeners' lives. Surely, some covet the limousine-lifestyle of Howard Stern, who enjoys a 23-million listener audience, and makes gazillions of dollars from books, movies, lunch boxes, and action figures.

Disc Jockeys

The job, in a real small nutshell, is to play music, read the news, sports, weather, traffic, and commercials during your on-air shift. Duh. When you're off the air, you also get to do production work. That is, read advertising copy for commercials, make a few voice-overs for public service announcements, or be a funny voice for the latest morning show bit. You'll be expected not only to read from prepared scripts, but also to make up some shit as you go along.

Don't expect to walk right in the door and get the morning drive slot. More than likely, you'll get to stay up all night playing music for third shifters, alcoholics, the unemployed, and psychopaths.

What you will actually do at the station which ultimately hires you depends entirely on that station's market. In this case, forget what your girlfriend always tells you—size does matter. The size of the station will determine your job duties and responsibilities.

At small stations, you'll give the corn reports, and you will read the news, and probably have to do some reporting. You might also double as the program director. You will go out to the hog auctions and tell the world what a durn fine piece of livestock Elmer Jacobson has. You will go to the regional meet between your hometown basketball team and the evil, gutless team from Durkins County. You will get to interview Melvin Carpenter, candidate for city council.

At the larger stations, especially where news jobs are concerned, your job will more focused and limited. Some announcers at large stations specialize in sports or weather. If they specialize in general news, they might be called newscasters, anchors, or news analysts.

Many radio stations mooch meteorologists from their television counterparts. Frequently, television and radio stations are owned by the same parent company, and then the lucky weatherman gets to pull double duty. At smaller stations, the weather might come from a wire service, or they might simply snap on the television and mooch it right from there.

A growing trend in radio is the use of news services, such as Shadow Broadcast Services, Medialink, or Metro Traffic. These companies provide news, weather, and other information content for radio stations. Radio stations love them because they don't have to hire news staffers and so they can save gobs of money that they would otherwise pay out in salaries. So, the next time you hear traffic, news, or weather, you now know that you might be tuning into a generic commodity.

Curious about those oddly repetitive traffic reports that you hear all the time? Our super-sly insider, who will tell her story only under a shroud of anonymity, gives us the inside scoop on traffic reporting: "The traffic gets thick at the same places each morning," she says. "A lot of times, the disc jockeys will just put in the tape from the previous day. After all, it's the same thing. It happens more than you think. Plus, if the helicopter pilot calls in sick or if he wants to take the day off, it's not uncommon to play a tape from the day before or even from last week. If they know they're going to be gone the next day, they might give another report and tell us to hold it until it's supposed to be broadcast. This is one of radio's big secrets."

Real radio sportscasters select, write, and broadcast the sports news. They might go to the field themselves to get the story or broadcast a game live on the air. At small stations, where the sportscaster position doesn't exist, this becomes the job of the disc jockey.

Talk show hosts

Talk radio burst through the roof in the mid-90s. Hosts such as Rush Limbaugh, G. Gordon Liddy, and Art Bell brought in millions of listeners to talk about things trivial and important. Talk show subjects can cover anything from heated political discussions to sports to stock market picks to love-life problems to automotive maintenance. Mostly, however, talk radio made its way with politics.

In *Mediaweek* (April 8, 1996) CNBC Equal Time anchor and former Bush campaign strategist Mary Matalin said, "Talk radio has definitely influenced elections. Forty-four percent of the people in the last election said they got their political information from talk radio. For many, it's their only source. For others, it's in addition to the information they get through the mainstream press."

The power and influence of talk radio is pervasive. It's relied on by people for news and information delivery. In February, 1996, Interp Research performed a study showing that, of those who listened to talk radio stations, 86 percent got their news and information from those same non-news stations.

Newscasting

Though not as glamorous or colorful as the big shot disc jockeys or talk leviathans (I don't know how many 18-year-old co-eds run screaming down to the radio studio to throw their moist panties at Hugh Downs), there is work in the newscasting realm of radio.

Überbroadcaster and charismatic newsman Paul Harvey has 23 million listeners who tune in daily on 1,200 radio stations across the fruited plain. Paul Harvey has been sending his confident, honest, Midwestern voice into homes throughout the nation for more than five decades.

As a reporter, you can't just stumble blindly into a news story. You have to know some background on the issues you are covering so you can ask intelligent questions and report what people need and want to hear. You have to be able to effectively interview the people who make the news and, to keep the story moving, you have to get the catchiest sound bite from them for your story.

"Newscasters need to be well-versed in the stories they cover," recommends University of Missouri professor and broadcasting researcher Vernon Stone. "This involves the expert use of information from periodicals, wire services and, increasingly, computer databases. Broadcast reporters bring together these elements into a concise and cohesive format that uses words and sound to communicate the event or condition."

The world under my thumb

And how about the money? Guys like Dan Rather and Paul Harvey and Charles Kurault have big, steaming stacks of cash just lying around their houses. They fill up their swimming pools with hundred dollar bills and backstroke through them at the end of the day. Could that be you?

"Sorry, but you'll have to forget the million dollar-plus salaries commanded by a few star anchors," says Dr. Stone. "Dan Rather started, like most did, at the bottom, earning minimum wage (then 75 cents an hour) at a small radio station in Texas in the early 1950s."

Don't let that dissuade you, though. It just means you have to put your licks in. Or you have to find a style that is so attractive and boosts the station's ratings so high in the sky that you can demand much more cash. Realistically, the average salary for a news announcer starts at about $31,200 but can range all the way up to $102, 676 in larger markets.

First you have to get in and jobs are always opening up. Either people leave the business or they climb the ladder of success, leaving a few rungs open at the bottom for...guess who?

According to the Federal Bureau of Labor Statistics, there will not much change in the field of broadcasting through the year 2005, due to the slowing growth of new radio stations. This means that there won't be a bunch of brand new stations looking for brand new voices. So, if you get a job out there, you're going to be replacing those who washed out because they didn't have the skills, those who couldn't stick out the crummy pay at the bottom, or those who did stick it out and moved up.

On-air reality check

Newscasters, announcers, disc jockeys, talk show hosts, and other "name" jobs are much more volatile positions than those held by the station secretary, promotions staff, or sales people. If you are interested in an on-air job, take this quiz to find out if you are well-suited for it.

1.   How often do you listen to the radio?
      a)   All day, every day.
      b)   Only when the OJ verdict was read, and when my tape deck is broken.
      c)   Why would I want to listen to the radio? They never play any songs I like.
2.   Hearing the same song over and over and over and over...

a)   Doesn't bother me at all, especially if it's some delightful, country song performed by somebody named Luke or Garth.

b)   Would start to nag at me after I heard it for the four-hundredth time.

c)   Would send me into a berserker rage, especially if it was "Horse With No Name" by America.

3.   How much money do you expect to make at your first radio job?

a)   Just enough to live on frozen burritos, ramen and Tic-Tacs, and sleep at my second cousin Purvis' farm on the outskirts of Asscrack, Iowa.

b)   Enough to have a new home, support a wife, child and a new—though not flashy—car.

c)   Enough to zoom to work every morning in my red Ferrari, leave immediately after my shift to meet some of the chaps at the yacht club, and trip off for weekends in Barbados a couple of times a month.

4.   I envision my work environment to be...

a)   All polished marble and glass, in ultra-modern studios. I, of course, will have my own private office and a gorgeous secretary.

b)   Stuff hanging together with chewing gum and duct tape until the next year's budget is approved by management.

c)   A mix of some old equipment that can still get the job done and some new stuff to at least stay competitive in the progressively more competitive radio game.

5.   My co-workers will be...

a)   Nuts.

b)   Well-read and well-versed on current affairs; witty and charming and always professional.

c)   A mix of a and b, tilted more towards nuts.

6.   The best time for me to work is...

a)   Mornings, when people are driving to work and need a little humor or charm to wake them up and prepare them for their day.

b)   Afternoons, when people are coming home from work and need to hear a calm, pleasant communicator who will help them unwind.

c)   Late at night when insomniacs, alcoholics, and psychotics need someone to blame for the problems in their lives.

7.   As part of a radio station team, my role would be...

a)   Their undisputed leader. Without me, they would all fail. Without me, there would be no radio station. Without me, there would be no way these deadbeats could eat.

b)   A small, insignificant cog in a giant machine who can and probably will be replaced.

c)   An important part of the team. I rely on others to do their jobs professionally. They also rely on me to maintain a high standard of professionalism.

8.   How well do you handle rejection?

a)   Great. I don't take it personally or get offended by someone who might not like some particular detail about me, such the sound of my voice, or some other intangible personal trait.

b)   Not very well. When I asked one girl to the prom, she said "no" so I locked myself in my bedroom for a week with a big bottle of vanilla extract.

c)   When I'm rejected, I get very hostile and like to tell people off. Sometimes, weapons are involved.

9.   Do you have a lot of perseverance?

a)   No. If the first place I send my resume to doesn't take me, I just quit. Is McDonald's hiring?

b)   Yes. Yes I do. If a dozen places don't want to hire me, I'll send a resume to the thirteenth and fourteenth place, and keep on going until I get the job I deserve.

c)   I'll give it a shot, but if I don't get a job in the first few months, forget it. I'll be back delivering pizza the next day.

10.   How do you like traveling?

a)   Great. Just so long as I get to go to a big city, like Los Angeles, Miami, or New York. Somewhere with a lot of culture and coffee shops.

b)   Traveling sucks. I want to work in one place my entire career.

c)   I like traveling. I'm especially eager to see Shawnee Mission, Kansas, and maybe even Moab, Utah.

11.   How important is job security to you?

a)   Very important. I want to work in one place for my entire career—preferably in a big city, like New York or Los Angeles.

b)   It is a big deal. If I got fired, I would take it *very* personally, and would probably move back in with my parents and take a job as a security guard at a grain silo.

c) Who cares? If I get shitcanned, what do I care? I'll just find a new place—maybe even a better place—to work.

Check your score!

Some of the answers were painfully obvious, but let's go over them one by one so you get an idea of the issues and logic out there. If you got any of them wrong, don't sweat it. We'll straighten you out.

1. How often do you listen to the radio?

a. All day, every day.

If you want to do any work in radio, you must understand the industry. What are the morning drive guys talking about? What kinds of ads are running? What kinds of sponsors do different stations have? These are all things you can learn only from spending time each day listening to the radio.

Plus, if you want to make a career out of something, it's best to do something you love. Do you want to be in radio just because you covet Howard Stern's limousine lifestyle? If you're just in it for the money, you'll have a rude awakening. People who are doing this job successfully started out in the business because they liked the medium.

2. Hearing the same song over and over and over and over?

a. Doesn't bother me at all.

It would be ideal if these songs didn't annoy you, but you're human. Things like annoying songs are going to gnaw at you. It will especially gnaw at you if you hate the radio station's format. A 20-year-radio veteran said, "Some of these songs drive me crazy. If it was up to me, I wouldn't listen to any of this shit. But it's not my call. I have to do what the program director says."

Is there a certain song that drives you nuts when you hear it? Does it burn a hole in your brain, forcing you to change the station, lest you go mad? At least, as a listener, you have the option to change that station. As a disc jockey, you won't be as lucky. Not only do you have to play the hot new country hit "Get Your Tongue Out of My Mouth 'Cuz I'm Kissing You Goodbye," over and over and over until your brain turns to a gooey mush, but you must sound like you genuinely like the God-forsaken thing.

"Easily fixed," you might say. "I'll just play my own songs." No dice, my friend. Remember, the station pays consultants huge stacks of cash to come up with what people want to hear. If you deviate from the list, kiss your ass good-bye. Disc jockeys have been fired for purposely straying from the play list. Welcome to corporate America.

3. How much money do you expect to make at your first radio job?

a. Just enough to live on frozen burritos, ramen and Tic-Tacs and sleep at my second cousin, Purvis', farm on the outskirts of Asscrack, Iowa.

I am sorry to report that life in radio does not start with a monumental salary. It makes you wonder why the job competition is so fierce. Only the people at the top make giant stacks of cash. Does that mean you'll never get a raise or get to move out of Purvis' basement? No way. Radio is truly one of those careers where the only place you can go is up. Not only will you make a teensy, weensy salary, but you'll probably start part-time. This means, if you want to survive, you'll have to get a second job somewhere at the beginning.

4. I envision my work environment to be...

c: A mix of some old equipment that can still get the job done and some new stuff to at least stay competitive in the progressively more competitive radio game.

This might not be an entirely fair question. Everything really depends on the radio station management and their commitment to equipment upgrade. Don't count on top-of-the-line equipment at every station, but you don't have to expect a studio full of crap, either. If stations want to stay competitive, they make sure the equipment works. But, if the station manager can save a buck, he'll keep the old microphones from 1956 in use.

5. My co-workers will be...

c. A mix of a and b, tilted more towards "nuts."

It's sad to say, but there are some nutcases out there. However, this works to your advantage. It also is part of the on-going quest for bigger and better radio jobs. An integral part of finding a radio job is networking, getting to know people, and letting them know what you can do.

Jim Bollella, a production director from Twin Cities, Minnesota, says you will run into people you used to work with. "People move around a lot. It's just the nature of the business," he says. "That can be a good thing, because if someone you worked with in, say, Nebraska, is the program director at a station you want to go to, he already knows you. He already likes you, you're not a gamble. And with so many wackos in this business, it's better to go with you, the sure bet, than to gamble on some nut."

Don't worry too much about the wackos. It doesn't matter if you are working at a radio station, at Burger King, or building space shuttles. The world is full of blockheads, but be aware that some of the more interesting freaks of the world do decide to get radio jobs.

6. The best time for me to work is...

c. Late at night when insomniacs, alcoholics, and psychotics need someone to blame for the problems in their lives.

Well, it might not be the best time for *you* to work, but chances are it is when you'll start. Get used to these terms:

- Graveyard shift
- Dog watch

I hope you like the work of the Commodores, because, my friend, at your first job there will definitely be some "sweet sounds coming down on the night shift." And the person putting those sweet sounds out will be you.

Sorry to be a downer, but *no one* starts in the morning drive spot. *No one* starts in the afternoon drive spot. Those are the time slots when the radio station is really pulling in listeners. They are going with the guy or gal who not only has been there the longest but has proven he or she can attract listeners.

Everyone gets broken in on the overnight shift. Not only are you likely to start at 11 p.m. and work until three a.m. or start at three a.m. and go to six a.m., but you probably will spend a few years in that slot. When you move from market to market, you might end up back on the night shift. If you move from the Dubuque, Iowa, station where you're doing evenings to Albuquerque, New Mexico, you might start back with the drunks and psychos.

29

7. As part of a radio station team, my role will be...

b. A small, insignificant cog in a giant machine who can and probably will be replaced.

As a piece of fresh meat, you can't afford a big ego. If you do have one, two things could happen:

• You get buried on the midnight shift until you are replaced;

• You get shitcanned.

There are fresh-faced radio school graduates out there looking for work, there are seasoned veterans who want to change stations, and there is the guy who just got fired because a new program director came in and didn't like his sound. So now would be a good time to grab a Bible or Bob Bennett's *Book Of Virtues* and brush up on your humility.

Even after a decade in the business, Minneapolis disc jockey Ken Ross says the last thing you can have is an attitude. "I thank God every day for this job," says Ross. "You never know when it will end, so I'm glad to have it. I appreciate the opportunity and try to do the best job I can."

Even if you work your way to the top of the heap, an ego can be your undoing. Depending on the radio climate, you can have the top show in the nation one day and the next day you're on overnights. Jim Bollella agrees. "I've seen people with the biggest egos get knocked off their perches," he says. "One day, you can be on top, the next day you're right at the bottom. It all depends on your ratings."

8. How well do you handle rejection?

a. Great. I don't take it personally or get offended by someone who might not like some particular detail about me, such as the sound of my voice, or some other intangible personal trait.

Okay, this was a loaded question. I hope, by now, you have caught on to the notion that you will face rejection before you land a job. There is more than a fair amount of rejection—*personal* rejection—in this business. This is a job where you "just aren't cut out for this work," or so says the program director of the fifteenth station to which you sent your demo tape.

This is a business where you sell yourself, not just a package of college education, the right internships, and knowing the right people. That's great for doctors and lawyers who spent most of their lives in school and know the head of the law firm or the chief physician. But in the case of radio, it doesn't work that way. It's all you.

9. Do you have a lot of perseverance?

b. Yes. Yes I do. If a dozen places don't want to hire me, I'll send a resume to the thirteenth and fourteenth places and keep going until I get the job I deserve.

This ties in with question number eight. The guys who give up after the first bit of rejection and resistance will have an impossible time getting a job. Remember, just because you graduated from broadcasting school, no one is going to be in a big hurry to give you a job.

On the contrary. You'll be out there fighting not only your former classmates for jobs, but also people with more experience. Here is where a positive attitude and making an effort come into play. It's easy to lie down and quit after a few tries. But quitting will not get you that job, will it?

Hell, no.

One disc jockey friend of ours says it's hard, but you must develop an immunity to criticism. "People call in all the time and say how much you suck," she says. "But, then again, that's their opinion. Other people out there like how you sound. Other people really don't care. You can't let the people who don't like you bother you. You just can't, or you'll go crazy."

10. How do you like traveling?
a. I like to travel. I'm especially eager to see Shawnee Mission, Kansas, and maybe even Moab, Utah.

The chances of your landing a radio job in the city you want to live in are right up there with getting struck by lightning while scratching off a winning lottery ticket as Big Foot walks though your backyard to greet visiting aliens. It's not going to happen. Remember, as fresh meat, you start in small markets reading the hog reports before Orville does the live report from the county fair.

But you can't get too comfortable in Podunk, Illinois, because, as we already mentioned, radio jobs are not the most stable in the world. Keeping or losing a job might have absolutely nothing to do with you or your talents. This is where the concept of networking helps you. It is very common for a whole radio department to get fired and, when one guy gets in at another station, to open the door for his buddies.

So, learn how to get along and play well with others. The dickhead newscaster you hate today could be saving you from stalking and killing deer with your bare hands for food tomorrow.

11. How important is job security to you?
c. Who cares? If I get shitcanned, what do I care? I'll just find a new place—maybe even a better place—to work.

This is the correct attitude to have because if you get into a radio career, you will get fired for no reason at least once, and probably more often. "There is a common saying people in radio have," says Bollella. "You haven't made it in this business unless you've been fired at least twice."

That is the cold, hard fact of the business. If you have spent any time around disc jockeys or talking with others interested in the business, this fact should come through loud and clear. "It's just a part of the business—a part nobody likes to talk about," says the Byrd, a former KSHE St. Louis morning man (*Billboard,* December 1994).

Realize, however, that everyone has to go through this. So what if you get fired? Everyone has been fired. You'll get fired. "Like any form of entertainment, this is a risky business," says the Byrd. "It's tough to say good-bye to friends you've made. But in order to survive, you've got to have a little bit of gypsy blood in you."

Mardit keeps a fairly brave outlook and positive attitude in the face of unemployment. It is an attitude that, to be successful, you must take on. "When you do what you do for the love of it, you never think of it as a job," says Mardit. "You stick with it through the hard times. I dodged the bullets for years and finally got hit. That's okay—I'll recover."

Give yourself one point for every answer you got right, and no points for answers you got wrong. Add those points up. It really doesn't matter what your total is, I think the point is clear. You must have a genuine interest in broadcasting to stay in it for any length of time. The money does not flow like water in a river at first. If you stick it out long enough, you start climbing the ladder and getting more and more cash. But you have to love it.

Producing

Do you like to juggle?

If so, producing may be your bag.

Producers are required to take all the pieces of a project's puzzle and find which pieces fit where and which pieces fit with which people. Kate Wood, a producer with Burkewood Communications, says all producers need to know how to do three things: communicate, manage, and motivate.

A producer gets an assignment to produce a commercial, for instance. The producer must determine what the client's needs are (and get totally and obsessively thrilled about those needs in the process). Then she has to build a team who can make that commercial come to life. From there, the producer oversees the project, making sure that all the little details are tended to, and making sure that everything is done correctly.

In fact, says Wood, a producer might find herself doing the teeniest, tiniest, niggling little details, for the betterment of the project. "If you believe that you can communicate, manage, and motivate well, and you can prove it in solid accomplishments in school or on the job, then you might have the makings of a good producer," advises Wood.

Radio production takes on many different faces. The work of someone producing commercials, comedy bits, and public service announcements is different from the work of producers who put talk shows together. At the core, the work is the same—organize, communicate, manage, and motivate. Beyond that, however, variations do exist. For example, when you're booking guests, you communicate differently from the way you do when you are dealing with radio staffers.

If you want to produce, expect to be paid around $20,000 per year to start. You'll find yourself doing all the *real* behind-the-scenes work. Producers are the ones who come up with show topics, they're the ones responsible for all the different attributes of the show. They are also people who, as I mentioned in the last chapter, must be good jugglers and deal with a million concerns and details at once.

24-7

If you are the producer, when the show is over and the intro music for the next program starts, *you will still be working*. After the last guest has left the building, *you will still be working*. While your show's host is dining at some Republican Party fund-raiser, showing off his new Italian shoes and chit-chatting about how vacationing at Martha's Vineyard is far superior to a week in the Hamptons, *you will still be working*. At night, when everyone has gone to sleep and late-night talk shows are going through their line of guests, *you will still be working*.

At 5 a.m., when your alarm clock goes off, you will start working. Good morning!

The life and work of a good radio producer never ends, and there are a zillion details. Details from the mundane (is the 9:15 a.m. ad in the cart machine?) to the enormously detailed (pre-interviewing an expert on tax and fiscal policy disparities before the radio show).

Larry Schwartz, a Boston producer, says in order to be topnotch, you must dedicate your life to the job. "The first thing on my mind when I wake up is the show, and the very last thing on my mind before I to shut my eyes at night is the show."

Schwartz starts his day by turning on his radio, TV, and computer simultaneously to listen and watch the news and check the wire services from home. He has been known to drive for miles to the town where news is happening and personally seek out guests (*Billboard*, Feb. 12, 1994).

Reading, writing, listening, watching

Successful producers don't stop working just because they're off the clock. If you are the type of person who (like me) hates working if you are not getting paid, then the life of a producer is not the life for you.

During the work day, producers research topics for upcoming shows, book guests, and keep their fingers on the pulse of the station's listeners. After hours, producers read absolutely everything they can get their hands on to stay current. They watch television and listen to other radio programs at the same time. The best producing stars read over a dozen newspapers every day.

The point isn't to rip off anyone's idea. (Although if you score a great show because you mooched off someone else, good for you. That show probably ripped it off from someone else.) The point of all this extracurricular research is to stay up-to-date on the issues. You have to talk about what people are interested in. You have to pay attention to current events and interests, and know the latest up-to-the-minute scoop. Plus, you have to be able to predict the future, a little. If it looks like a big news event is about to happen, you have to be there. And you have to keep up with the on-going issues that your listeners have shown interest in, just in case something comes back to haunt you.

It's not just a matter of reading the latest *Time* magazine and discovering that the world's fossil fuels will all be expended by 2050. You must absorb that, find out if it is a topic your listeners might be interested in, and get a show together that's appropriate for your station.

First, you read everything under the sun on the topic. You do as much research as you humanly can do. Then, you line up some guests, preferably pro and con on the issue. Next, you do a brief pre-interview and find out each guest's take on the topic.

Picking guests is a challenge. You will decide if they are "radio worthy." If they have crummy voices or aren't very good conversationalists, you better go back and find someone more suitable. If they suck, it's your fault. Not theirs. Certainly, it's not the host's fault.

Then, you take all the information you've found and brief the host. (It would be silly if the host had to do his or her own work.) Some good, professional hosts will do some research on their own, but it's your job to make sure they know what they are talking about before they go on the air and make total asses of themselves.

Now, you get the guests either on the telephone or in the studio and hope the whole thing works out.

Frankly, I think the job would suck. Sure, you get to be the heartbeat and the lifeblood of the program, but it works the other way, too. If something goes wrong, people blame you. Plus, if you want to be the best at this job, you wind up putting in a ton of *unpaid* hours. Screw that, man.

On the up side, if you're "super organized" and really like to be in charge of a project, this job is for you. If you want to do a really stellar job, you must work like a Japanese beaver day and night. Maybe you like that idea.

"I look forward to two-hour traffic backups," says Larry King producer Pat Piper (*Billboard*, Feb. 12, 1994). "The longer the back-up, the better the show. It gives me time to listen to the radio and find out what people are talking about. That's what makes a successful show." Piper says he starts his day by reading six newspapers, 25 magazines, and a stack of tabloids he calls "essential."

Let's not forget about technology. For all you techno-gearheads out there, hooking up to the Internet is important. Getting on-line gets your finger right on the pulse of the nation. I'll admit joining the "hot and horny cheerleader" chat room will not do much for the next day's broadcast, but joining a chat room where the participants are talking about real issues (not just what color panties all the women are wearing) can be a great source of questions, public opinion, and even background material.

Emotional problems

Though producers do enjoy a fast-paced-live-on-the-razor's-edge-of-death-leave-a-pair-of-clean-underwear-and-send-a-corsage-to-my-wife lifestyle, the job doesn't come without its down sides.

"On the day the Challenger blew up I found a guy who worked with the shuttle program, so I got him in the studio," said Piper. "As he was telling his story with tears rolling down his face, I just thought to myself 'what kind of beast am I?' He devoted his life to these people, and here I am putting him in front of a million people to talk about it. I never felt so bad." (*Billboard*, Feb. 12, 1994)

The job does have its light side, too. While thinking about the Challenger staffer, Piper also remembered the day he booked a Soviet journalist to speak on the radio show.

He didn't find out that she didn't speak a word of English until she showed up with an interpreter.

"Producing is probably the hardest job I've ever had, and it's the most humbling, too," says The Gil Gross Show producer Gregg Cockrell. "A lot of times you don't hit a home run, and you feel it and your host feels it. Well, I just feel so stupid. I feel like a failure. Yet, at other times I feel like a genius."

In addition to all the reading, researching, television watching, and radio listening, producers take on one more thing. Some know it as "The Thing That Came From The '80s." It's that beast, that hideous, pop culture, yuppified word you see in employment articles and books. It's that word that (although dreaded) is so important. It's that word which describes a process no one knows how to do effectively, but everyone says you have to do.

It's "networking."

In the life of a radio producer, networking goes beyond trying to make that jump to another, better paying job. Networking, in this capacity, refers to tracking down authoritative, interesting guests who will have something productive and positive to offer your program. You never know which contacts will garner viable, well-regarded guests. Maybe a chat with a heart specialist will lead you to an ear, nose, and throat guy who can talk intelligently about the new strain of strep throat savaging your listeners. Networking (again, unpaid work) helps you build your expert base, will make your show better, and make you more valuable to the station.

Okay, that sounds great and you may be ready to say, "Hey, Bobb, I'm ready to start my networking list right now, dammit." And that's great. If you are the type of person who can schmooze and make these contacts, great.

However, if you aren't a "people person," you will fight an uphill battle as a producer.

Radio producers must be able to speak intelligently and personably with the guests. They have to be able to make the interpersonal connections that garner great results. They have to be "out there" schmoozing and meeting people and getting involved in all kinds of things. So if you are a perennial wallflower who can't stand to deal with people, you should think twice about pursuing a career producing a radio show.

Further, be aware that producing jobs are not the same at every station, on every program, or in every market. The life of a talk show producer differs greatly from the producer of a Golden Oldies show. Both are faced with the same basic mandate—assemble all the components to make a good show. The route each must take, however, is much different. If you are producing the oldies show, you go to the music library and pull out all the Del Shannon and Bobby Vinton compact discs. If you are producing "The Day in Review" or some such current events crap, it will be your responsibility to track down guests, stay topical, do your research, and all the rest.

Sounds exciting, right? Wait up. Even though you are doing the majority of the work, as a producer, you get paid next to nothing at the outset. I know full-time producers at some radio stations who still take part-time jobs to pay the rent and keep gas in the car. Take heart though. As you climb that mystical ladder, you will make more cash. When you start working for the Howard Sternses and Don Imuses, you can quit one of your night jobs.

Sales

Remember Herb Tarlek from *WKRP in Cincinnati*? Remember the cheesy suits and matching white patent-leather belts and shoes? Welcome to radio sales!

Actually, nothing could be further from the truth. At least that's what radio salespeople say. No matter what career I had, if the poster boy for it was Herb Tarlek, I'd raise a fuss, too.

Paying the bills

Everyone hates radio commercials, but they serve a vital purpose. They keep the radio station on the air. The revenue brought in by radio sales not only covers your future paychecks, but also capital procurements like digital this and satellite that. Further, the sales provide funding for operating expenses—where do you think the money for caller number 26's free boom-box comes from, anyway?

Today in American radio, there are more than 65,000 sales people out there pounding the pavement to generate more than $2 billion in national radio revenues. Especially in markets with heavy competition, sales can be a cut-throat, fast-paced job.

According to a 1997 survey conducted by Edison Media Research, 45 percent of advertising executives rank radio as the most effective way to reach young consumers and most effective in repeatedly reaching all consumers. Radio sales will continue to flourish as a way to advertise.

What does this mean for the ad rep? Sales, sales, and more sales. Ad sales that equal big paychecks.

Though some would shy away from any sort of sales job, radio sales can be a silver bullet to station ownership, management opportunities, and hefty salaries. Of course, you do have to take a few licks on your way up the radio marketing ladder.

George C. Hyde, executive vice-president of the Radio Advertising Bureau, says that radio sales at every level are virtually the same. "Identify the marketing goals and needs of prospective advertisers, develop a radio marketing plan to meet those goals and needs, convince the prospect of the plan's value, and monitor implementation of the plan to insure success," said Hyde.

You suck! Don't ever call here again

Sales is one of those careers where you need thick skin and perseverance to succeed even *after* you land the job. Throughout your career as a salesperson, you must continue to make the client happy.

As an entry-level salesman, you will do a lot of cold calling. You'll end up calling owners and managers of small and medium-sized businesses in your area. When you finally get someone to let you through the door, that's when you have to convince them to buy time on the station and fork over a pile of their cash.

Sounds simple? Not so fast. You'll have to do more than take the owner out for a three-martini lunch and a golf game. You will be expected to research the type of business your prospective client operates, then develop a sales strategy that will convince the client that yours is the best radio station to handle his needs.

Hyde advises salespeople to be good listeners and questioners. "When you get across the desk from the prospect for the first time, your emphasis should be on asking questions and listening, not talking," says Hyde. "The more you learn about a client's business needs and aspirations, the better you'll be at developing a successful radio marketing plan which addresses and satisfies those needs and aspirations!"

Did I mention that you'll work on commission and not get paid for any of this work until the advertiser signs on the dotted line and the check clears the bank?

I am sure you have excellent sales prowess. I am sure you can sell g-strings and pasties to nuns. However, your responsibility does not end when Tool World signs on for a dozen spots a day.

As an Account Executive, (that's what radio stations call the sales crew), you have to make sure the clients stay happy by making sure the ads sound the way they expect them to sound, or better. Also, you have to make sure the client wants to keep sending in more checks to re-run the fabulous, profit-producing ads.

Remember, you'll get paid based on the amount of money you rake into the station. The pay structure for sales staffers varies slightly from station to station. At some stations, sales reps earn a base salary plus commission, if they're lucky. Most places pay you straight commission. But the good thing is that the longer you stay in the business, the easier it will be to make money, and you will make more and more of it.

Expect to earn about $20,000 at a small station, and $50,000 at a big one. Don't let that frighten you. Once you get rolling and build up a clientele, making those sales will be easy. You'll get to know what your clients want and how to deliver. But you can't get complacent. Advertisers fall by the wayside now and then, and their places must be filled.

Hyde warns that if a client stagnates too much, you can be replaced on the account by another salesman who can get the money pumping back into the station. After a few successful years as a radio salesman, Hyde says, you'll be likely to have three opportunities for career advancement.

1. Local sales manager's position at your station;
2. Sales position at a larger station in the same area;
3. Sales position at a station in a larger market.

I'd pick number three because the bigger the market, the bigger the sales! That's right, more money for you.

If you are up for it, a radio sales job may be your best bet to get in the door at a station. Stations are constantly on the prowl for salespeople who can go out there and bring in the bucks. In other words, sales jobs are available. At smaller stations, often the disc jockeys and announcers double as salespeople.

It's a good ladder to climb, too. Some salespeople start wearing managerial hats while still in the trenches of larger, networked stations. The jump up from sales person to manager is predictable, if you're good. Local, regional, and national sales manager jobs are often filled by current sales people who have worked in the trenches.

You don't even have to have a college degree, a high school diploma, or a library card to be a salesman. Will it help? You bet it will. Experts recommend a degree in marketing or advertising, and plenty of courses in English, and broadcasting.

Sell air time and your soul

God bless anyone who can be a salesman and not go nuts. I wouldn't want to be a salesman. Truth be told, I couldn't sell firewood to an Eskimo. I couldn't sell oxygen to the crew of Apollo 13. I couldn't sell panties and a bad rug to Marv Albert. But I am not you. You are you, and you might be damned good at it and end up with a career in radio making more money than anyone you know.

These days more than ever, a sales job requires a knowledge of radio. Customers are more savvy. Sure, there are dopes out there who will buy radio ads but have no idea if you ran the ads or just took their money and bought a new entertainment center for your house. But most business people are going to be pretty hip to the system and you have to be, as well.

You will want to kill

The hard part of the job seems to be getting all those ads sold, right? Actually, no. The tough part of the job comes afterwards when you have to keep your clients happy and maintain the flow of checks.

Advertisers want to see results. If Big Dave's Muffler Shop buys a few ads from your radio station, Big Dave expects to fix more mufflers. If the ad doesn't bring him more customers, he will cancel it. Wouldn't you? At the very least, he will want an explanation from you, the one who sold him these ads, as to why his sale on Teflon-coated carbon-filament mufflers is not the blowout you promised.

So, you beg him to give you another chance. Then, rewrite the ad. Develop a new concept, change the writing, and recording. Add more special effects and edit it better. At each and every stage of ad development, expect your clients to question you. This micro-management will not happen in every case, in fact, most ad campaigns will come off with no problems whatsoever. But some clients will piss you off.

Now, you get paid, right?

The part of the sales job that turns most people off (a lot of them quit the same week they start) is getting paid on commission. Sometimes, you don't get your commission until the client's check clears the bank. This can lead to some pretty lean paychecks. However, if you can stick it out through the scrawny paychecks, you could do very well at the end.

Take a look at some salesmen who have been in the business for many years. From the twelfth hole, holding a cellular phone, the guy can make more money in one day than the rookie salesman can make all week cold-calling potential advertisers. It might not be fair for the poor sucker calling the local birdseed store trying to sell 30-second ads, but the guy on the golf course has paid his dues. He started where you're going to start. You'll get there, eventually.

Promotion

When you think promotions, think sales. Promotions is really an off-shoot of the sales department. The promotions department not only sells the station, but keeps advertisers happy.

Think about the last fair or festival you went to. You probably saw every radio station in the area out there with some goofy setup giving away plastic bags or paper hats or some other crap. The DJ was probably broadcasting from some huge, inflatable boom box.

Though it might seem like a simple project, getting your disc jockey into an inflatable boom-box and giving away sunglasses with the radio station logo emblazoned on the side is not the easiest task in the world. Or when the radio station gives away tickets to the Madonna concert, it's not just a matter of the disc jockey calling up Madonna and saying, "Hey, Sweet Cakes. How about a few front row center seats for our listeners?"

No, no, no. Promotions projects take many radio staffers to put them together. So you want to be in the promotions department? It is a career in station's sales department that doesn't involve old-fashioned selling or commission.

Spin city

"Your job is to create an image of the station," says Chris Monroe, a former promotions director in St. Paul, Minnesota. "Whatever image that station has, it's your job to project that. You have to personalize the station for each listener."

Don't let the lofty notion of winning the hearts and minds of listeners all around town fill your head with undeserved self-importance. "It isn't rocket science," adds Monroe. "You just need the ability to ask, 'What do the listeners want?'"

According to Monroe, the ex-promoter, hers is a nit-picky career. "You have to be very detail oriented," she says. "There are tons of details, and you have to be able to anticipate and take care of every one of those details."

Usually, a promotion starts with an advertiser saying, "I want something cool tied in with my product." Sounds simple enough. Your client sells baseballs, so you set up a deal with the local baseball team to have Free Baseball Day at the stadium.

"It can be really tough," says Monroe. "Clients want us to do fun things with their product or with their service, but it just doesn't work that easily. It's hard to think up a good promotion centered around Dixie cups. But that's the kind of thing we face."

Let's go back to our state fair example. Not only did Monroe have the problem of promoting the radio station at an event, but also competing against a dozen or more other radio stations at the fair. "We had to sit down and ask, 'What is it that people want at the fair?'" explains Monroe. "Then it occurred to us. People want a place to sit down. So our whole promotion was geared around giving people a place to relax and take a load off."

To accomplish this, Monroe had to arrange for the extra space on the fairgrounds; she had to find tables; arrange security; get big water jugs; set up signs; generate on-air announcements; and manage to make it all work.

Though she loved her job, Monroe admits that the job of promotions director is extremely time consuming. During major events, it is not uncommon to put in 12-hour days. (Incidentally, you'll be salaried, so don't get too horny for that overtime cash.) "You have to do it," says Monroe. "You absolutely have to stay on top of the work or they'll find someone else to do it."

The right stuff

Beyond developing an event, a successful promoter has to be mindful of marketing, and how to make the details of a promotion work best. "You have to know where the banners should be placed, and what to say in the press release, and what the ads should say," advises Monroe.

What kind of person becomes a promoter? "Most people end up here because they want to help," Monroe says. "They think it's cool, it's fun. And that's a quality they need. They have to see this as a fun thing."

It's important to have an outgoing personality, and not to be afraid to take charge of a situation to get the job done. In terms of educational background, Monroe recommends a communications and marketing degree from a four-year college. "That kind of education is good to have," she says. "You have to know how to type a letter and be able to communicate effectively."

On the other hand, even a promotions pro such as Monroe says you that can get her job without a four-year degree. "It's not absolutely necessary, but it's always valuable. It means that you wanted to learn more so you can do a good job."

A vital part of a promoter's job is the ability to communicate. if you did not develop this skill in high school, you must get hold of it before you get into radio.

"You have to know how to write. It's very important," says Monroe. "There are a million memos that have to go out. You're talking to a million people. You have to do it successfully."

As the bottom-rung promotions assistant, look forward to a salary in the $25,000 to $30,000 per year range. It goes up from there.

The ugliness that is promotion

"Promotions directors usually explode, implode, or end up going into sales," says Chris Monroe, the *former* promotions director. When you think about listing high-stress jobs, this one should be on top.

Of course, all the weight will not be on your shoulders when you get your first promotions job. You will start as a lowly assistant, and then work your way up from there. As you climb that ladder, you must take advantage of opportunities to learn everything you can about promotions because, when you get to be the boss, all the shit will fall on your shoulders. Of course, by then you'll know what to do.

Promotions is, without hyperbole, the most detail-specific job in the entire world. "You have to juggle a million things," says Monroe. "You really have got to be detail-oriented, and that can be the tough part for a lot of people."

Working in radio promotions keeps you on the go. No sitting around, twirling your thumbs. If you are in even a little over your head, you're going to suck a lot of water into your lungs. "I have whole days on my calendar filled up," says Monroe. "There are some days when all I have time to do is run down my list and make sure everything gets done."

The details are where you can sink, swim, or get eaten alive by the sharks. For the most part, Joe Q. Public won't know if everything in your promotions plan went off without a hitch. If item #1,768 on your list is "get paper towels" and it never happens, don't fret. Probably no one noticed. However, there are going to be some extremely important entries on your list, and you can't skip them.

What's for dinner?

One promotions director who coordinated a pizza party for a bunch of screaming kids and their parents before a "Disney on Ice" program forgot one thing.

A few hundred parents with their screaming kids showed up at the arena for a pizza party before the big show. The promotions staffers thought they had everything covered.

- They had enough staffers for 700 people.
- They had the place settings all made up, complete with personalized name tags.
- They had the free gifts lined up for all the kids.
- They had Mickey and Pluto and the rest of the Styrofoam-headed freaks set up to entertain the kiddies before the show.
- They had matching T-shirts for all the staffers.
- They had banners and buttons and confetti and hats with Disney logos all over them.
- They had door prizes and free popcorn coupons.
- They had the radio station's logo everywhere.
- They had every kind of soda pop known to man.
- They even had gourmet coffee for mom and dad.

They had everything a pizza party needs...except for the pizza!

Only once everyone was seated, and Mickey Mouse and Donald Duck had left the banquet room, did promoters realize that no one had made arrangements for the pizza. A call to several local pizza shops revealed that, in fact, no one would be able to accommodate pizza for 700 at the last possible second.

There is a *somewhat* happy ending to this story. Unfortunately for those expecting a pizza party, a large pepperoni with mushrooms and onions was not an option. Instead, promotions and kitchen staff had to empty out a freezer, stoke up the ovens, and prepare an emergency dinner with the only thing available: Frozen cod.

In the rush to take care of every conceivable detail on her 2,763 item list of things to do, the promotions director did not remember to take care of item number one. Order Pizza for 700 people. Of course, that promotions director never had to worry about making such a stupid mistake again. She got shitcanned.

Still, a job in the promotions department can be fun.

About 80 percent of promotional campaigns are initiated by the advertisers who want remote broadcasts from their Oldsmobile dealerships or by the advertisers who want your listeners to hear "Back to the '70s" night with a heating-and-air-conditioning focus.

This is where your creative thinking comes into play. It's not just theoretical creative thinking, like dreaming up a subject for your dormitory discussion group. There has to be a definite, successful end result. This great, creative idea has to be able to satisfy the client, and make the station look like the greatest radio station in the entire world.

"It can be really, really hard," says Monroe. "Most clients think what they have is so important that we should do a special promotion for them. Unfortunately, most of their products turn out to be stupid and lame, and we end up having to do a promotion for something ridiculous."

You can turn this problem of having a lame advertiser into your big opportunity. After all, you have to please the client. They pay your salary, basically. So take that piece-of-garbage of a product or service and turn it into the greatest promotion the world has ever seen!

How might I serve you?

The people in the promotions department do not get the respect they deserve. Most radio people regard the job as some sort of weenie-assed sissy service job. Unfortunately, they just don't get it. Where do they think their money is coming from anyway?

The whole radio shebang relies on advertisers (unless you work at a public radio station, and you have to pretend that your so-called grant-makers and sponsors are not advertisers while you help yourself to everyone's taxes). Everyone at a radio station has a vested interest in keeping the sponsors happy. That's all there is to it.

In the promotions department, you get up close and personal with the advertisers. You're right there where the money is, while the on-air folks are a step removed. So, sure, disc jockeys must be respectful and not say stuff like, "Here's another stupid ad from Eddie Beckmann's used car lot. Did I tell you they screwed my buddy when he needed warranty work done on his car? Man, do they suck."

You're fired on the spot if you say that, of course. However, disc jockeys really don't care much where the ads come from or if one company is replaced by another.

Even the advertising department, which is responsible for kissing advertisers' asses and thanking them for their frequent checks, gets to pass the buck off on the promotions department from time to time.

Let's face it. Sometimes, advertisers with utterly non-promotable products or services want a promotions job done for them on your radio station. "There is a huge, huge, burnout ratio," says Monroe. "You can find yourself one day in a parking lot next to a huge inflatable oven saying to yourself, 'What on earth am I doing here?'"

Glamorous, on the surface at least

You know those hot tickets to see Eric Clapton, Janet Jackson, Sting, U2, Nine Inch Nails, and every other big name act that radio stations give away? Guess who arranges those free tickets? Guess who arranges the free T-shirts and the limo rides? Guess who arranges the back stage passes?

That's you, my friend, the promotion director.

Most people probably think that promotion directors and radio station bigwigs are always hanging out with the band after a big show, nailing groupies, and drinking 40-ounce bottles of malt liquor with Dr. Dre.

I am sorry to report this is not the case.

After a recent visit with Dr. Dre (we talked about 'hoes, why Tha Man is out to keep a brotha down and why free trade agreements with Mexico and Canada are actually a detriment to workers, both in America and abroad, and should be nullified through pending Senate legislation endorsed by a bi-partisan House and Senate commission) no one got to hang with the party people after the big show.

The truth is, promoters exist to make sure everything goes off without a hitch. "Most people think it's fun and games," says one promotions staffer. "They don't have a clue how much work, how much real, honest-to-God work it is to line these people up." From getting the celebrity's "people" to commit to the appearance, to hammering out all the little details, and making sure that all the celebrities are happy, you'll have your hands full.

Think back to all those stories you've heard about prima donna celebrities who demand six not seven, not five, but six bottles of Evian water; five pounds of M & Ms with the brown ones picked out; and exactly 123 white towels in their dressing rooms. You'll get to deal with this kind of bullshit.

"People would say it's cool to be back stage at a big concert. People would say it's cool to work on one of these projects. And, yeah, it is kind of cool, but it gets real old real quick," Monroe comments. "It's an enormous amount of work."

So what if you get burnt out one day? You can still have a great time and a very cool job until then. You'll have stories to tell your grandchildren. You'll have a more exciting job than all of your friends put together. If it sounds good, then the promotions department might be just the place for you.

Engineering

Every radio station has to have at least one licensed chief engineer to keep the transmitter transmitting, the swear switch switching, and the thousands of blinking lights blinking. The chief engineer's responsibility is to inspect, repair, and maintain the station's equipment.

Though the planet is full of cave men and women who think technology is the worst thing ever concocted, the truth is that most electronic gizmos and gadgets work as intended. In fact, thanks to improved technology, there really isn't a need anymore for full-time engineers. So most stations contract out for the job. In other words, the tech-head comes walking through the station for maintenance and inspection, but he doesn't live there.

The only reason they need this guy at all is for routine maintenance and if, unlikely as it might be, all this grand, infallible equipment goes on the fritz. In that case, the guy ends up with a phone call in the middle of the night to degauss and recalibrate the gonkulator, or some such thing.

In addition to the chief engineer, each station must also have a licensed operator who monitors the transmitter while the station is on the air. Usually, however, this job ends up in the lap of the disc jockey.

Me, I like an electrical shock now and then

If you like gizmos, gadgets, flashing lights, and a 60 cycle per second hum, radio engineering is your bag. From its early days, engineering has evolved from radio tubes and huge bulky chunks of monophonic broadcasting equipment to today's digital age of compressed data streams and mux units.

The more dorky and nerdy you are the better. The further your pants slide down, revealing the crack of your ass while you recalibrate the DBX-1000SFX to be in synch with the BROADSAT uplink unit, the better. And, like other geeky, soldering-based jobs, it helps if you have never been stark naked with someone else.

In the control room of a radio studio, technicians operate equipment that regulates the signal strength, clarity, and range of sounds of recordings or broadcasts. They also operate control panels to select the source of the material. Broadcast technicians in small stations perform a variety of these duties. In large stations, and at networks, technicians are more specialized, although job assignments might change from day to day.

The terms "operator" and "engineer" and "technician" are often used interchangeably to describe the various radio engineering jobs. Transmitter operators monitor and log outgoing signals and operate transmitters. Maintenance technicians set up, adjust, service, and repair all the electronic broadcasting equipment. Audio control engineers regulate sound pickup and transmission. Field technicians set up and operate portable field transmission equipment for broadcasting outside the studio.

Obviously, as an engineer, your radio training is going to be different from everyone else's at the station. You'll be taking very technical courses, and you'll have to be able to figure out how to fix stuff and keep stuff running. Everyone else at the station needs to know only that the equipment works. You're the one who has to know *why* and *how* the system works.

What dumb people will do

Don't worry about people skills. You don't have to wear a suit and tie or carry an alligator briefcase to the office every day. In fact, the most conversation you'll need to have is to tell your boss you fixed the problem or what part you have to order. To deal with everyone else at the station, learn these handy phrases:

- What color was the smoke?
- What did the smoke smell like?
- Tell me again why you dumped the Sprite on the control board.
- Next time, call me. Don't jam a paper clip into the cart machine.

Yeah, yeah, yeah... How much do I get paid?

As a contract engineer, you can expect to work for several different radio stations at the same time, which means that if you have a boss, if you get the job done and everything is up to snuff, he shouldn't be in your face too much.

The Bureau of Labor Statistics tells me that the average technician at a radio station earns $28,000 a year. For chief engineers, average earnings can go higher than $57,000. There were roughly 40,000 engineers in 1996. The numbers have probably gone up since then.

The job of an engineer is completely different from the other careers listed in this book. Whether the differences make the job better or worse is for you to decide. Though I like the idea of tinkering with gadgets and computers and wizzits and whatnots, the job might not be for everyone.

It seems as if every 10 minutes broadcasting equipment changes, requiring technicians to keep up with the latest and greatest in hardware and software. If you are the type of person who loves new electronics, engineering is your field. Keeping up with new technology and loving it will make you an outstanding broadcast engineer. However, if you are the type of person who wants to learn it once, then live off that knowledge for the rest of your life, this wouldn't be the best job for you.

Broadcasting has kept pace with technology. As radio tubes became obsolete, so did the schmucks who refused to learn what a transistor does. When microchips replaced transistors, a bunch of technicians were shown the door and the new blood came in. To be a successful engineer, you have to keep up with technology. The best way to do that is to genuinely love the field.

Another thing that separates engineers from the rest of the radio pack is results. A disc jockey's job is more subjective. There's the voice quality. And then there's the personality. It's hard to measure the day-to-day results of a disc jockey or a producer's work. Sure, you can go by the ratings, but it's still awfully subjective.

Engineers, on the other hand, are easy to measure. If the signal isn't going out, obviously the engineer screwed something up. If there is a weird hum in the background, you don't have to wait for the Arbitron ratings to come out to know the equipment isn't working properly.

Some people like jobs with an absolute result. Either the equipment works correctly or it doesn't. No one can really argue with it. On the other hand, you maybe the type of person who prefers shades of gray. In the disc jockey's job, some will like your work, some won't. But in the world of engineering, there is no "It sort of works."

I hate math. I hate it with all my heart and soul. I can't add, I can't subtract, I can't multiply, and things like factors scare the ever-living shit out of me. But if you are a math dork, engineering is your thing! You get to use a bunch of different little formulas to figure out the resistance ratio across a circuit or why the capacitance value in the mother-board is off.

Math is a game of absolutes. Either the answer is right or it's wrong. The answer is not open to interpretation. If you're a left brain sort of person, run—don't walk—to the nearest vo-tech school and sign up for your future job in broadcast engineering.

Beepers, three a.m., and you

Remember when pagers came out? You always thought guys who had them—like obstetricians—were important. Now any *schmuck* can get a pager, and most people do. In fact, I think I'm the only person left in civilization who doesn't have a pager.

As an engineer, you will probably get a pager paid for by the station. Why would they be so generous? Because, when some little gear falls off its shaft at three a.m., the overnight disc jockey needs to get some poor sucker (you) on the phone to un-fuck the radio station.

Things rarely break down between the hours of nine a.m. and five p.m. Clip this handy list out of the book and keep it with you at all times. It will be help you figure out when the equipment will malfunction.

- Your second day on the job at two a.m.
- Christmas Eve.
- Christmas morning, right while you're unwrapping your presents and drinking your Bloody Mary.

- Right when you're about to get laid.
- At the best part of an action movie.
- When you're sleeping off a four a.m. beer blast.
- Ten minutes after you've been back from fixing the equipment the first time.
- Whenever it's 30 degrees below zero outside.

So what's the upside to being on call? Nothing, really. Being on call sucks. If you are salaried, you won't even get paid more when your pager goes off. Moreover, you're going to be paged by stupid night-shifters whenever a light bulb burns out, or some fool unplugs the CD player from the outlet, or when the toilet clogs up. On the plus side, if you are paid by the hour, every time your pager goes off you snag some overtime.

Engineering for a radio station is not a bad gig. Usually, you won't have a bunch of uptight corporate office types breathing down your neck, and you can pretty much act yourself. Plus, fun does happen at radio stations and, when it does, you might be there.

Production

Do you have a favorite radio ad or morning show bit that makes you laugh your ass off every time you hear it? If you have ever wondered where those bits come from or ever thought you'd like to create something like that, then production is right up your alley.

Production staff—not to be confused with producers—are the ones who take the concept for an ad or an on-air bit, break it down into little pieces, then reassemble the pieces into the final, finished, on-air form.

Someone had to do it

Those clever song parodies that Rush Limbaugh plays are not things he just slapped together in his basement to fill some air time. When your favorite morning show host plays one of those oh-so-wacky bits that they think are so hilarious, but are usually pretty stupid, that's production.

Though they account for just a few seconds or minutes of time on the air, production spots take many hours of time to think up, write, record, edit, re-record, and put together.

Jim Bollella is a production director for a major Twin Cities, Minnesota, radio station. "We write commercial copy, put on-air bits together for the morning show, line up voice talent, and put it all together," he says.

Production engineers also are responsible for those little on-air quickies you always hear, but never think about. After all, someone has to write a script, find an announcer, record the announcer a few dozen times as he tries the line a few different ways, edit it, and put it onto a tape that the disc jockeys can play when they need to fill five seconds.

Someone is responsible for making that tape that always says, "You're listening to the best new rock, KRCE."

Depending on station size, the duties of these producers vary, as in most radio jobs. At small stations, the on-air personalities do more of the creative work. They might think up clever bits or write the script for a song parody themselves. Spots at large stations, on the other hand, might be generated by public relations or advertising firms. "In my smaller market days, I did a lot more writing than I do now," says Bollella.

Of course, production engineers do not have to work for a radio station alone. There is plenty of work out there for someone who is interested in production, but who does not necessarily want to punch a corporate clock. During a break in radio station jobs, Bollella worked for a production house. "It was literally in someone's basement," he recalls.

Train me

Though you do not *need* a college degree to be a production engineer, Bollella, a former college professor, highly recommends it. "It's rare for anyone to be in this job for long," says Bollella. Plus, he adds, many companies will be more likely to hire a production engineer who has some sheepskin. "There are a lot of companies who will hire you if you have a four-year degree in anything," he said. "If you have that degree, you can get in the door almost anywhere."

Production engineers are tinkerers. Tinkerers don't like to get their hands dirty. They mess with the small stuff. Rather than using hammers and nails, production engineers use other tools to put together advertisements and things like, "This is a public service announcement from your friendly Heartland Chevy dealer."

"You've got to be a thinker," says Bollella. "You have to try things different ways. You have to make sure they sound perfect. You have to try something and, if it doesn't work, try something else. You have to develop an ear for what sounds good. But don't worry, it happens."

Bollella says the talent might not come overnight, but when a production engineer develops his "ear," it can be a great asset. "I really do equate it to a musician's ear," says Bollella. "You can say 'I love the way that sounds,' and not know a thing about it. But a musician can listen to it and say, 'actually, that's just a three-chord piece of garbage.' But it doesn't matter. You know when it sounds right."

Your main tools will be recording equipment, scripts, and computers. "The software we are using is getting more and more complex," says Bollella. "You have to know how to use computers to do this job."

Pick up those computer skills during your college education, Bollella recommended earlier. "Creative writing skills are also important. Sometimes people who want to work in production don't understand that you have to be able to write creatively to do a good job," he says.

Once a piece is written, the on-air personalities do the most evident part of production work, the announcing. "We have a guy we hire to do all the promos for the station," says Bollella. "By using one guy, the station gets a voice. When people hear his voice, they know he's talking about our station. But for other things, like ads, we use the jocks."

Other on-air talent is hired when musicians or singers are needed for a project. "A lot of work goes into these projects," said Bollella. "It's not like we can just slap this stuff together on the spur of the moment."

As a production engineer, you won't have a high public profile. You don't have to sell anything to anyone. What I am trying to get at here is that, if it matters to you, you can pretty much dress like a slob when you do this job. The only ones who will see you are some of the disc jockeys, the program director, and anyone who sees you on your way to the bathroom after drinking coffee all morning.

Production engineers usually start somewhere between $25,000 and $35,000 per year.

I'll do it my damn self!

Though the world of production is thirsty for talented, creative producers, they won't take any *schmuck* right off the street. As is the case with all radio careers, you will start at the bottom. Producers not only work as staffers for radio stations, but they also can perform freelance work for different radio stations.

Freelancing is a nice business, if you can get the ball rolling. Every aspect of the job is on you. You have to buy your own equipment, and you have all the responsibility to consistently produce quality projects. On the happy side of this coin, you are your own boss. No one orders you around. No one demands that you punch a clock. Freelance work is all you. If you screw up, it's your fault. If you succeed and become the hottest property in the nation, it's all to your credit.

To prepare for production, a college degree from a good school is highly recommended. History, English, art, music, and other liberal arts studies will help you along the way. Studying liberal arts is important because it will broaden your horizons and bolster your creativity.

As a producer in the fluid, hired-today-fired-tomorrow world of radio, you have to be able to work in a station with any format. If you develop a commercial for "F'd Up Malt

Liquor," it might not play very well on the local classical station. So, and pardon my political correctness here, you have to diversify yourself.

Life as a production engineer can be a ton of fun, especially if you like to tinker. This note a little higher, this laugh a little longer...you get the picture. A good production engineer takes a lot of tiny things and mixes them together to make a solid, effective piece of work.

Making it all make sense

The longer you work in the business, the easier the job will become. Of course, this is the case with any job, so don't rest on that handy bit of advice. Take a minute and roll through your radio dial. For once, try to avoid stopping at a great song or a zany talk radio brain freeze about the latest sex scandal. Instead, find a commercial.

Now listen to it. Break it down into teensy pieces. What elements are present?

- Music
- Dialogue
- Maybe dialogue from a second or third announcer
- Sound effects
- Finally, that last guy in every ad saying, "Visit your local burger joint this week at one of five thousand handy locations and win a free flat soda with your greasy bag of fries."
- Mixing it all together

Each element in that ad had to be put together in such a way that you, the listener, are not only encouraged to buy a burger, but also to think the radio station pulled it off as consummate professionals. Now, let's look at what went into each of the elements of the ad.

Music: Someone had to either track down and obtain the legal rights to a certain bit of music, or hire someone to write an original jingle and pay the musicians and singers to perform the thing.

Dialogue: Someone (maybe an ad rep, but probably you) was asked to write the witty dialogue between an imaginary couple, Stan and Rita. He or she had to fit cheesy terms like "summer clearance" "hot deals" and "bargain basement prices" into a normal, everyday conversation between a married couple.

It might sound easy, but advertisers and ad reps have certain, ear-catching phrases they want you to work in to the ad. Did I mention it has to fit into a 15-second spot?

Special effects: This is pretty easy. There are probably a zillion different sound effects in your production computer, but if you have something special in mind for a certain ad you must go out in search of that certain, special sound effect.

Putting it all together: Now comes the really dorky, geeky, chess club part of the job. It's probably the part you can have the most fun with, too. You get to take all those puzzle pieces and make them fit together perfectly.

Is the music coming up at the right time? How about the dialogue? How about that legal crap at the end? Can anyone understand it? Do the sound effects sound right or are they out of place? Do the sounds come in too early, too late, too loud or too soft?

That's just an quick explanation of some of the technical details you'll be dealing with whenever you produce a spot. As you work on each one, your ear will get more and more confident. Eventually, people will be asking you for your expert opinion and hiring you to produce their ads for them. At that point, you can start demanding some serious cash.

2. Landing a Job

But I've lived here all my life

To New York or not to New York, that is the question. If William Shakespeare had written this book, I imagine he would start this section with that meaningful dilemma in mind. However, since he is not writing this book (in case you didn't know, he has been dead for several years), I'll take some liberties.

For many careers, such as book publishing, advertising, or biochemical terrorism, the best way into the business is relocation to The Big Apple. Let's face it, the action is there. Madison Avenue, the big networks, the movers and shakers are there. Is New York the place to start your radio career? Hell, no.

In fact, in the radio world, cities such as New York, Washington, DC, and Los Angeles are the ultimate goals, the pinnacles of success. They are for big shots in the business, and the big-bucks radio careers. The entry-level radio jobs in the big city are few and far between.

Location, Location

There are millions of young people eager to get away from their small town roots. Have you heard that practically every child raised in Nebraska grows up and moves away? These ambitious youngsters can't wait to follow the latest, greatest fad and the big-time job opportunities to New York or Los Angeles. For some career paths, it's the best move. But for radio, it's the kiss of death.

In the radio business it is simply a fact of life: You must start small. You must start in the sticks. You can't walk into a big, hot New York radio station and say, "I want to be on the air at this station." In fact, you can't walk in at all.

So, let's hope that you love the small town atmosphere, that flannel-shirt, Twin-Peaks feeling of pie on the coffee shop table. If you do love it, at least temporarily, radio can be the best job for you, at least from a location point of view. On the other hand, if you hate your parents, if you hate the United States government, and you can't wait to get out of small town USA, chances are you'll have adjustment problems in Podunkville, Indiana.

The typical ladder of radio success starts in a small town and ends in a major market, such as New York, Chicago, or Los Angeles. Of course, you can stay in a small market if you want to, depending on your personal preferences.

"I hate big cities. I don't ever want to set foot in a big city as long as I live," says Marc Jackson, a disc jockey in small-town Nebraska. Marc doesn't rake in a stack of cash every year (around $14,000), but he has a career he loves.

For many people, small-town life is appealing. There is a simplicity of life there. You get to know everyone. Maybe you'll run for Mayor one day, who knows? The cold, hard, impersonal city just doesn't do it for you. You crave a simpler, white-picket-fence-type life.

However, there are career advancement issues for such a person. In a small town, if you choose to stick around, you will probably get bumped up to better and better time slots, first of all. Ultimately, you'll wend your way into the program director's job or some other managerial position. If you see yourself on this path, you should stick around and do your best, busting-ass at your first radio station to make sure they don't get rid of you before they get used to you.

Take advantage of a good thing

Of course, you might want out of the small-town market as quickly as possible. Make up your mind to tolerate what you have. Pretend you can stand it for as long as it takes to get the whole gig over with. Here's what you have to do to make that initial small-time job work for you.

Suck it up at first. Chances are, you're starting as a part-timer. You're either driving miles upon miles to work or you've relocated to Palookaville. Take every chance you can to get into the studio, learn from the other people, and hone your skills. This is your chance to develop your show.

This also is a good time to get involved with the station and the network. Get to know as many people there as possible, and learn how they do their jobs. The more you learn and can do, the more marketable you will become.

If you are a disc jockey, go in and help the engineer maintain some equipment, help the broadcast engineers put together some radio ads. I'm not a big fan of working for free, but don't look at it as if you're giving away something for nothing. Look at it as learning, and the price you pay is a few hours of your time.

71

If you're in school, get involved in the events. If you can broadcast the local high school's football games, your name becomes familiar to the community. The idea here is to do everything you possibly can to build up your knowledge of the business. For God's sake, if you do wind up in a small town, don't waste your time. Don't just watch the clock. Use every opportunity to better yourself.

Bear in mind, if you have one of those chains connecting your nose to your ear piercings, it might not go over very well (they probably would have picked that up in the interview anyway). Consider dressing like a normal human being while you're at work. Then later, on your own time, if you really feel the need to go to an artsy-fartsy coffee shop and listen to pretentious poets reading about how awful their lives are, just get in your car and go. Screw the planet. Burn as much fuel as you can to get yourself out of that little town for awhile.

For those of you who do love small towns, this whole notion of making a name for yourself in the community and developing your skills is just as important as it is for those who can't wait to run, screaming, from Durkin County, South Carolina. If you want to stick around, you have to prove yourself worthy. That means things like professional development and community recognition.

First, the better your job skills, the more valuable you will be to the company. However, remember that radio jobs are not the most stable income sources around. So, when you get fired (and you will), the skills and contacts that you bothered to build up will make you more competitive in the job market. The goal is to develop the best audition tape you can, and constantly improve it. That way, when you do get fired, you won't be up shit creek.

From a standpoint of growth and development with one employer, the more valuable you become to them, the more cash they are going to throw at you. This will be good because, at this point, the generosity of Uncle Cletus and Aunt Beulah is probably running out and you'll need to find your own pad. Plus, it's hard to bring chicks home when Uncle Cletus is yodeling along with his favorite Slim Whitman album from the moment he gets off the tractor until he passes out from drinking corn squeezins at night.

That satellite took my job

In this day and age, even those small-town jobs are harder and harder to come by. Technology isn't just putting brewery workers and assembly line workers out of their jobs. More and more small-town radio jobs are gobbled up by automation. "Automation" is the dirtiest word you can say to someone in radio. It means the whole show is coming off of some damn satellite. Unfortunately for those of you starting careers as disc jockeys, your new job might get taken over by a robot.

The situation gets even worse when you consider that automation developers are making the evil machines more and more appealing to radio stations every day. Even radio stations in direct competition with each other can subscribe to the same automation service and have subtle, touchy-feely differences built in.

"You can now customize each radio station and sell to multiple outlets in the same market," says Jay Batista, sales manager for radio automator Wegener Communications (*Broadcasting & Cable*, Oct. 3, 1994). Technology allows the automators to customize and fine-tune their service.

On the plus side, automation provides more work for engineers. When the XP-9000 DJ-A-Tron breaks down, someone has to fix the thing. Plus, you get to go in there, change its batteries, put in a new PCV valve, and rotate its tires, or whatever maintenance you have to do.

Sales careers are not as threatened by automation as are disc jockeys and broadcast engineers. After all, someone still has to make the money to pay for the satellite service.

Hey, no way is some technology going to replace my distinctive and irresistible personality, you might say. No way will anyone fire me from any station once they find out how great I am. Major markets, here I come! It's just a matter of putting in my time.

Calling all future radio personalities! This is a lesson you have to learn right off the bat: Don't get a big head! No matter where you are, humility is an extremely important quality to possess. Think back to names like Shadoe Stevens, Rick Dees, and Wolfman Jack. They had their time in the sun. They were on top. Now, they're nothing but names in a Trivial Pursuit game. No matter where your career leads you, you can't be self-righteous and arrogant.

"There's always someone else out there better than you," says Allen Phillips, a former morning man in Iowa. "I know it's a cliché, but it's true. Just when you think you're top dog, someone comes by and pisses on your head."

That fact is every bit as true in Ely, Minnesota, as it is in Chicago or New York. You can get fired from a small-town job as easily and unpredictably as you can at the Number One Station in the Universe. It's cool to be number one, but have a little class. Don't get cocky.

Cash on the dollar

Just for fun, you can compare your current salary in your area with salaries in other areas. On the world wide web, go to www.homefair.com/homefair/cmr/salcalc.html. This site has a form that asks your current location and salary. Then, based on where you are moving to, it will tell you how much the same job would pay. Kind of fun. Might be depressing. Sometimes, too much knowledge is a bad thing.

The Easiest Way In

Without a doubt, the absolute best way to get a radio job is to sneak in the back door by finding some easy, no-brain job at the radio station. Network your ass off. Then find yourself pulling the overnight, Sunday morning, or fill-in on shifts. This is absolutely the best, most sure-fire, top-secret way into a good radio job.

Now you know. This is the page that has the answer. I am sorry, but there is no magic fairy dust you can sprinkle on yourself or on the program director. There is no chant you can intone while you rock back and forth in the station lobby. Doing entry-level, menial work at the beginning is the best and possibly only way to get in.

It's not so bad, though. Doing grunt work at a radio station does a couple things for you.

• First, you get to meet people and you are not perceived as a threat to anyone's job; and

• Second, if you have a question or want to help out one of the disc jockeys, they will probably be willing to help a poor part-time non-threatening person like you.

Best of all, as a beginner, you learn the inner workings of radio. You learn the dynamics of the people who work there, what the company honchos want, and if you aren't a total cock, the disc jockeys will probably give you some tips for success.

Then, if you can properly schmooze the program director, he will put you on the air. At the very least, he or she can give you some educated, informed, and free advice on your demo tape.

Does all of this mean you can start out at some big-ass station and, within a year, be challenging Howard Stern?

Hell no. It means that if you get in the door, you'll have a good chance of getting on the air. And, if you can wrangle it, you might be able to start at a pretty decent station where you aren't projecting hog futures for the rest of your life. Unless you're into that. If so, then please move to Belle Plaine, Iowa, right now. Thank you.

Wayward soul does well

Donna, a kindly, but horribly misguided friend of my wife's, spent a kazillion dollars on law school and, despite my best efforts to bring her back to the side of good, is now a lawyer. However, because of a part-time job she does just for the fun of it, she could probably get a real radio job if she wanted it.

When she's not plying her destructive legal trade on unsuspecting citizens, she works at a local big-wig radio station in the promotions department. At the events and fairs, she is the one who passes out the free buttons, the posters, and the other promotional clap-trap that people immediately throw away when they get home.

Donna has garnered a lot of respect at the station and could, if she wanted, start her own radio career. But she must have eaten paint chips as a child, because her goal in life is to be a lawyer. What a waste! On the radio side, she's doing more than most graduates of shake-and-bake radio schools ever do. On the legal side, she is probably earning a living, but why bother? Here's your lesson for the day: Do what Donna's doing, even though she's wasting her life as a lawyer. Get your foot in the door in any way possible.

Sure, they are not giving her her own morning show, but she's learning the business from the inside out. She knows the suits. She knows where the swear switch is on the console. In other words, she's there.

For those of you who want to make a career change, this is a great way to get in the door. Again, consider our friend, Donna. Although she's a sleazoid lawyer, she could be making a great career change. She grabbed a part-time job, just for fun, but she still earns enough to live. You can keep your day job and still get into radio. This is the route many have taken to get in the door.

Remember, radio doesn't stop on the weekends or at night. It's not a nine to five business. Get in any time you can. Work around your schedule. Once you get in, start asking around for some help and guidance. You can get the program director or a kindly disc jockey to help you out and cut a demo tape.

This notion of knocking at the back door is especially useful if your current job is one that has any contact with a radio station. For instance, if you are a salesman for office products and one of the stops on your route is the local radio station, a chit here and a chat there with the sales director could bring you a new job. Read what our friend Ken Ross did when he decided to pursue a radio career.

From researcher to radio

Ken Ross, a disc jockey already mentioned in this book, had no training whatsoever in radio. Now he is the evening man at KSTP-FM radio in St. Paul, Minnesota. "I was a researcher," explains Ross, who happened to be performing market research about radio listening preferences. "That was my foot in the door. I let my boss know I was interested in broadcasting and he helped me put together a demo tape."

Ross says job hunters who don't want to go to college, really don't have to bother. "If you think you have the talent, get your tape together, and send it out," says Ross. He also recommends networking to get your foot in the door. "Volunteer at a radio station, then let the boss know you are interested."

It took Ross ten years to get from Cincinnati to the Twin Cities. "I was fortunate," said Ross. "I don't take any of it for granted—you can't in radio. Even if you get a morning gig somewhere, you can find yourself out of a job just like that."

Ross advises broadcasting hopefuls to be optimistic and steadfast. "The breaks are out there, if you are persistent."

Racism and Sexism

I have good news for women and minorities and I have bad news for the evil white men who want radio jobs. Racism and sexism are alive and well in the radio industry. Therefore, stations need to meet quotas and quotas usually mean death to the white guys. In one case, a Peoria, Illinois, radio station was fined $10,000 by the FCC for failing to hire any minorities for its thirteen "hiring opportunities."

If you are a woman, a black person, or a member of any other minority groups, you should expect some resistance in your job quest. At least, so say a bunch of insiders and a series of fines handed down from FCC. On the other hand, stations looking to "get theirs," might snatch you right up because you have dark skin, a vagina, or both.

The good ol' boys club

"I am *so* glad you asked that question. People have to know that prejudice exists in this business," a black disc jockey told me. Though he said I could use his name in this book, I've decided to withhold it, not wanting him to suffer any repercussions.

"I've been pretty fortunate, but I'm also aware of the fact that I have to work extra hard," our whistle-blower says. Though cognizant of race problems, he says he doesn't let the issue consume him. "I'm not one who uses race as a scapegoat. No matter who you are, you have to work hard and do the best you can. But I do feel as though I have had to work harder than some others," he admitted.

For women in the broadcasting trenches, opportunities are not fruitful. One major-market female disc jockey, who asked to remain nameless, told me, "Women have a lot of opportunities, because the stations need to have a certain number of women. But that doesn't mean we get the jobs we should."

Beyond the question of "will I get a job or won't I?" there are other issues for women and minorities. Where do they stand in the station hierarchy? Which shifts will they be assigned? Sure, there are the Robin Quivers (if you're reading this book, you must know who she is) and the Ken Hamblins (a black *conservative* from Colorado) out there, but for the most part, women and minorities don't find themselves as the headliner in the morning show or doing afternoon drives.

"This really bugs me," says the unnamed female disc jockey. "There are shifts that women don't ever get, like afternoon drives. When you hear about a woman in an afternoon drive, you have to take note of it." Moreover, she says, a woman rarely leads the morning show. "They are always the sidekicks or the butts of the joke," she says. "Can you think of any woman who has her own morning show? I think this is still a guys' business. Guys get the better shifts. Every program director would say that's not an issue, but it is. It's sexual discrimination."

Broadcaster Joyce King agreed in a recent *Billboard* article. "Most minorities and women remain scattered, usually relegated to the over-night and late-night duty, and more often than not, paid pennies on the dollar compared with white males in the business," says King.

Some industry bigwigs say the issue isn't a matter of discrimination or bigotry, it's a matter of learning how to work and play well with others. "To a certain extent, it's a boys' club," says Terry Avery, the vice president of adult programming for Radio One in Maryland. In *Broadcasting and Cable* magazine (October 1995), she says, "If you know how to hang with the boys, you can get in the club." Avery adds that while some men have made patronizing or aggressive comments, coping successfully is a matter of style. "I deal with it. I don't get it every day, but sometimes somebody comes along who wants to strong-arm me."

Some entrepreneurial types have seized this opportunity to make their own business out of it. Compliance Surety Inc., of Colorado Springs, Colorado, started its own placement service for minorities. "People in radio said their biggest problem was communicating with the minority organizations that can supply qualified applicants for jobs. Our service will significantly reduce the time and expense involved in advertising and distributing information on available employment opportunities," says Conrad Naef, president of CSI (*Broadcasting and Cable*, Jan. 1, 1996).

Yeah, so what?

I've given you both sides of the coin here, with the promise that you can use strained race and sex relations to your benefit. According to some, racism and sexism will keep you pinned down to overnight shifts or relegated to very early Sunday mornings. According to others, quotas and fear of the evil overlord of the FCC can get you in the door. It's up to you.

If I were in your position, I would look up the most recent articles in *Broadcasting and Cable* and find out which discriminating bastards have been fined recently. Send a tape and resume and watch the job offers fly. From the standpoint of getting in the door, you should definitely play the race card. Use quotas to your advantage, and do whatever you have to do to get that first job. Then, once you get in, you can be Superfly, Shaft or Xena the Warrior Princess, if you want. You'll get fired and treated like shit, just like any other snot-nosed beginner.

The fast track to success

As a white man who has been screwed out of many jobs, scholarships, and God knows what because of affirmative action, I hate quotas. But, as the guy telling you how to land that first radio job, I have to whole-heartedly advise you to do whatever you can to get in the door. The hardest part of getting into any business is breaking in. Once you get in, your merits and skills are going to take you wherever you want to go.

Avery has the last word on this issue: "The bottom line is, we're all here to do a job and the one who does the best job wins."

Join the Army

Join the WHAT? You heard me, troop. Join the Army. Join the Navy. Join the Air Force, or the Marines. Christ, join the Coast Guard, for all I care.

I am absolutely the last person on the face of the earth who would advocate joining the US Military (except the Air Force where they have hot and cold running chicks).

I feel uniquely qualified to bash the military. I was in it. I was an Army Intelligence Specialist during the Gulf War, then I joined reserve officer training corps during college. Wisely, I dropped out of the program before I graduated and had to become an officer who screams at other people all day to buff floors.

Some think that enlisting serves a great and noble cause, that we're fending off the commies or keeping sadistic bastards like Saddam Hussein in check. In actuality, my peacetime service to Uncle Sam consisted of continually shining my boots, dusting the light fixtures in my room, and disassembling and reassembling five-ton trucks for no good reason.

Do more before eight a.m.

If you are a person of better moral fiber and character than I am and you are able to get up at five a.m. and run 27 miles without bitching or moaning, then maybe a career in any of our armed services is right up your alley. You'll do more before eight a.m. than most people do all day.

The military is a microcosm of the real world. The only difference is that everyone in the military has to wear a name tag. Plus, in the real world, no one teaches you how to shoot a machine gun or throw a hand grenade—at least until you go to public school. You can enlist for *any* job in the military. You can do anything from building bridges to typing letters to fixing or flying airplanes to cooking meals to blowing up enemy aircraft carriers.

NOW HEAR THIS: You can even get a job as a disc jockey or radio engineer. They train you for free, and when you're done, you'll have experience and a demo tape you can send out to land *real* jobs. Plus, if you opt for the GI Bill or some other college program, you'll be able to afford more training when you get out.

Not all roses and honeysuckle

Here are a few items you must consider before signing your name on the dotted line, and getting your head shaved, and your nuts (if you have them) probed. There are some rather important details the recruiter will not tell you about.

If you sign up as a broadcaster or announcer, make sure that's what you're signing up for and make sure it's guaranteed. The recruiter can make every promise under the sun, but unless it's in writing, it's not going to happen. The day after basic training, you could find yourself running next to a tank in South Korea. Often enlistees think they're joining up to be a broadcaster or a cook or whatever. But the Armed Forces have a funny way of keeping you from doing what you signed up to do.

I enlisted to be a military intelligence analyst, which I was, but only during the Gulf War. The rest of the time, I changed oil in trucks at the motor pool, cut grass, and cleaned my room. Also, I got to march in circles and see how quickly I could put on a charcoal-filled chemical warfare suit.

Under the guise of training, you will often be kept from doing your *real* job, the one you enlisted for. Also, during your military training, if *someone* decides you suck as a broadcaster or as an engineer, you will be reassigned "in accordance with the needs of the service." That means you get to run next to a tank in either Guam or Turkey. They don't tell you where you're going until right before you get on the plane!

I am the world's laziest man. The only time I run is to the fridge for another beer during a commercial. I thought this was no big deal when I enlisted, because I expected a desk job. I thought only the guys who drove tanks and blew up bridges and performed hand-to-hand combat had to run.

Think again, man. No matter what your job is, you will wake up at five a.m., you will run until you puke, and you will do a shitload of calisthenics. Even if you are a radio guy.

Why do they keep treating you like a soldier if you are just a cook or a disc jockey? Because, everyone is a soldier. At any time, your unit could be shipped to a jungle or a desert or some stupid island someplace where you'll have to start killing people. A lot of dolts who enlist don't even comprehend that it can happen. Remember the Gulf War, when a bunch of people suddenly decided to become conscientious objectors when they found out they would be going to the desert? They cried, "I only enlisted for the college money!" Where did these boners think they were? Disneyland? They joined the Army, for crying out loud. This is the *war* business.

Then, of course, there's the whole subservience thing. You have to call people who are complete morons and below your contempt "sir" or "ma'am." If you can't handle this, please do not enlist. Also, I hope you like working where everyone wears the exact same clothes. Day in and day out. No individuality is allowed.

If you must...

Yes, I'm tying to discourage you from enlisting. However, for the sake of fairness and to give you the big picture, you should know the options. One option is to get some free radio training by getting a job in the military.

Red flag: If you have any reservations about enlisting, don't. Four years is a long time to be miserable.

Once you're in, you can't get out. (Unless you want to pretend you're gay, then you're out in a heartbeat. Of course, everyone in camp will beat the shit out of you with a sock loaded with rocks in the toe.)

Do some virtual pushups on-line

You can visit the various branches of the military on the Internet:
Army: www.army.mil
Navy: www.navy.com
Marine Corps: www.usmc.mil
Air Force: www.af.mil

3. <u>Education</u>

Cramming your head full of knowledge

Fundamentally, there are two schools of thought on the need for higher education. Those schools are the "yes" school and the "no" school. Simple, right? It's not that simple, really. The "no" people are really not saying that you don't need any education at all. (No one is allowed to say that these days.) The division in the conceptual battle over educational preparation for a radio career really comes down to a debate about what kind of education is best.

Some people say a four-year university program is the best route. Others say, "Screw that. You only need to go to a broadcasting school and you're set." Of course, there are compelling arguments on both sides of the debate, and ultimately, it comes down to what you want out of your career and your life.

You should make the best start that you possibly can, and that start begins—you guessed it—in high school. I hope this book has gotten to some of you still pushing pencils in the local senior high. If so, great. Read on.

If you're already out of high school, bear with us while we get the minors up to speed here.

Corn-fed education

If you are reading this while you're still in high school, then *now* is the absolute greatest time to start sculpting your future. Don't worry, I didn't talk to your parents, guidance counselors, grandparents, or your overachieving cousin Alfie. Cliché time: I want you to get off on the right foot. So if you've been badgered by your parents about "do your best in high school, blah, blah, blah...." until your ears bleed, just put all that out of your mind.

Before you have to start putting out big bucks to educate yourself, you should get all the education that you can get for free. Get the basics.

The Basics

The guy who wants you to land that radio job is telling you something. Do well in school and succeed. Right now, you can get an incredible jump on your future broadcasting career. You have to pursue basic and, whenever possible, higher-level education in a gob of different fields to prepare yourself for eventual stardom.

Taking as many of these courses will give you a good base on which to build your creativity and skill: English, history, science, computer science, speech, debate, math, geography, political science, foreign languages. Even shop class is a good one, if it's electronics. After all, you're going to have to know something about how the hell the radio equipment works.

Screw phyical education. Sure, it's great to waste an hour playing basketball or volleyball with your buddies, but it won't help you get a radio job. I'm sure there are some whiners out there who complain about "health this" and "overweight that." But hey, baby, this isn't a health book. If it makes you whiners feel better, go run a mile and do 100 push-ups every night and feel as superior as you want.

High school is a great place to get a nice, even academic base established. Depending on your future radio career goal (in sales, on-air, engineering, or whatever), you'll want to emphasize one area or other later on.

You can, however, get some good early training in the trenches, too. Obviously, if you want to be an announcer, you should take as many speech and debate classes as you possibly can. A good understanding of math is fine, but you won't be needing "Advanced Pythagorean Geometry." On the other hand, if you want to be an engineer, try to be the first one in line to sign up for that class.

Go for a mix. Geography, history, English. You know the drill. Some of them might sound like rubbish, but say, one day, you find yourself reading a news story about Kuala Lumpur. You should be able to flash back to your high school education, and remember that it is not in Japan.

On-line, all the time

A solid computer education is always a good idea, but it's especially important for the future broadcasters among us. All aspects of radio station life have computer elements to them. Believe it or not, disc jockeys don't just plug a compact disc into a player and press the "play" button. Discs are routed through computers.

Producers, engineers, production engineers, and even the guy who keeps track of which songs the station plays use specialized computers and software every day. More and more stations are relying on them for all kinds of functions. If you don't know how to use a computer, then finding a job is going to be a colossal pain in the ass. Don't just focus on an Internet web browser or using some word processing program. Learn how the computer *works*. Learn something about hardware. Learn what an operating system is, and how to use several different applications. Go through one or two programming classes.

After school activities

As if a full day at school isn't enough, Dr. Bobb is here to plan your afternoon and evening for you, too. Ready?

Join the speech team, debate team, and the electronics clubs. If your school doesn't have these clubs, seek out the teachers in charge of these subjects and tell them you want a radio career and need their help. Most teachers will not only be glad to help, but they won't be able to pry the smiles off their faces when they find out a student actually wants to do something.

Don't talk yourself out of joining the speech, debate, or electronics clubs. Just do it. If there is something standing in between you and joining one of these extra-curriculars, figure out how to get around it.

Say your work schedule at Chicken Hut is so tough that Eddie, the 20-year-old assistant manager in charge of the deep fryer, won't let you join an extra-curricular. Then quit the job! It is very important to build the best base you can for your career.

You don't want to end up working at Chicken Hut *after* you graduate, do you? Give up the spare change now, in exchange for something better down the road. Also, all of these extra-curriculars will look great on your college applications, your resume, and will give you some excellent starting skills.

Get a job, man

You don't have to be a disc jockey or a technician to get a job at a radio station. You can get one now, during high school. You might have to do crap like emptying waste paper baskets or sweeping floors, but you will expose yourself to the equipment and the people involved.

They aren't hiring? See if you can volunteer. No one ever turns down free help. Tell the program director what you want to do with your life and see if he will help you.

You might wind up as an intern (unpaid gofer). But guess what? Now you have work experience for your resume.

Trade School or College

In the world of post-secondary education, there are two ways possible to spend your money. You can either shell out approximately $7,000 to a trade school or a bigger shit-pile of cash to a four-year college or university.

A trade school is not necessarily a better or worse choice than a four-year program. Each has its advantages and disadvantages. Personally, I think a four-year program has more pros in its corner, but you can still be a success after going to a trade.

Your decision should come down to a matter of filling your particular educational needs. Generally, trades give you a quickie, very focused education, while the four-year schools are supposed to make you a better person.

Shake-and-bake trade schools

"You can get a degree in as little as six months!" So say the commercials for the eight trillion trade and technical schools out there offering a radio broadcasting course.

To the obnoxious die-hard education crowd—you know, those people who think you need to smother yourself with Plato and student loans—trade schools are at the bottom of the barrel. They are places where lazy people go to get a quick, meaningless education. Pay, in; learn a little bit, out. You know the drill.

The courses generally run from six months to a year. A few of them are more detailed and hands-on and can take as long as the regular four-year university programs. These schools lack prestige. For some people out there struggling to get a radio job, the mere mention of a shake-and-bake school brings snickers and sneers. But, if you can stick out a little derision and name-calling, you can find yourself behind the microphone at an honest-to-goodness radio station inside a year. Of course, that microphone will likely be in Bumfuk, Iowa, but so what? You're in.

Trade schools do what they're supposed to do. They get you in, teach you the bread and butter of what you need to know to land a job. Then they give you a piece of paper with your name on it and their name on it in bigger, fancier letters.

Though a trade school diploma traditionally elicits a bit of a groan, you should be aware that technical schools are gaining more and more prominence and prestige. Studies are showing that a four-year college education is not necessary for everyone. In fact, trade school might be the best or the only affordable choice for you. After all, doctors can pay off their student loans by scaring old people to death on the table. You won't have that kind of fun. You'll be grinding it out on small change for quite a while, even if you do land a job right away. Don't go deep into a debt that will crush your head.

Since trade schools prey on the naiveté of young people who want to be the next Howard Stern, you have to be careful. Make sure the school isn't in the league of the "International Radio School of Poughkeepsie" or some other sleazy "pseudo-school." Make sure the one you pick is accredited by the National Association of Trade and Technical Schools, or by a state board of education.

There are gobs of radio hopefuls who go to these shake-and-bake schools. Program directors' desks are swamped with demos and resumes from trade school graduates. If you go there, be warned. The "education" portion of your resume isn't going to delight and dazzle, no matter what the trade school recruiter says.

Trade schools do, on the other hand, have one great selling point—free job placement. Through the school, you can land your first job. However, that job might not be at a radio station. You could end up disk jockeying for "Men in Uniform Night" at the gay bar downtown.

A big con for trade schools is not what you learn, but what you *don't* learn. Trades do not prepare you for life outside your chosen career field. "But what do you do when your radio career is over?" asks Jim Bollella, the production engineer in St. Paul, Minnesota. "You don't hear a lot of 65-year-old disc jockeys out there, so you have to take care and make sure you have something to fall back on. You need to have a Plan B."

For Plan B, build up your educational background so that if the bottom does fall out of radio, or you decide that broadcasting isn't for you, you're not stuck with nowhere else to turn. If you go to a trade school, you aren't going to have a Plan B.

Some night when you're out and about, try this. Stop into a local club or somewhere else with a disc jockey and ask the guy where he went to school. I'll bet you a dollar most of them wound up going to a trade school and are waiting for their "big break" into radio.

I'm not saying trade schools suck, but we have to be realistic. They get the job done, but probably not in the best way possible. Of course, if you're chomping at the bit to get out there and get on the air, by all means go to a trade school. Look in your yellow pages under "Schools," and sandwiched between the Academy of Aviation and the Drafting Academy are the broadcasting schools.

Keggers, loose sex, and student loans

These days many people think that a "well-rounded" individual has to go through a four-year degree program. If you do decide to take the snooty high road and go for a four-year degree, get out your wallet, because you are going to pay. Of course, I should probably be saying that to your parents.

There are two great big advantages to going to a four-year program and shelling out all that cash for college.

• Prestige. Radio stations will be a little more impressed with your training. It will be easier to get that first job offer. Also, having a college degree makes you a better candidate for advancement at the station, especially into management, if that is your goal.

• Better education. This is not to say the education you get at the six-month microphone academy isn't any good. But you'll get more at a four-year school.

In addition to whatever courses you can take that are even remotely connected to radio, you want a broad, well-balanced set of other junk to "round you out." Huh? The Accrediting Council on Education in journalism and mass communications advises you to learn about government, political science, economics, history, geography, at least one foreign language, English literature and composition.

So, from a how-much-time-did-I-spend-learning-radio-stuff perspective, you probably are only earning a quarter of your credits in mass communications and the rest of them in liberal arts studies.

Choose me, choose me....

Not all schools are created equally, so do yourself a favor. Check out the schools you are considering before you sign up, transfer any credits, or send any money. First, make a list of each school you want to attend and could probably get accepted to if you applied. Then consider the following:

• The school's training opportunities. Pick a school that uses modern, up-to-date equipment. You should be training on the same kind of equipment that you're going to be using when you land your first job, not on the dean's old garage-hobby stuff. (Some schools, believe it or not, use imaginary training equipment. That's great, if you're going to get the overnight spot in the Land of Make Believe. Maybe they'll teach you how to fly, too.)

• The professors. Try to find a school that employs real-life, honest-to-God radio professionals to teach the courses. Contrary to the old adage "Those who can, do, those who can't, teach," the best teachers are the ones who have been out in the field, getting their hands a little dirty. I had one professor in college who barely had a college degree but had lots of experience. He was teaching right alongside the jerks with masters and doctorates. I liked him because he knew what was going on and he was a better prof than all the brainiacs with advanced degrees.

The guys with experience are still going to use those wordy, philosophical textbooks. They have to. However, they also bring the "real life" or "in practice" view to your education, and that's precisely what you need. You don't need someone spouting all this rhetoric and jargon about the masses. You need to know how to do the job and how the big picture affects you. Professors with experience can, and will, help you a lot with the real-life portion of your education.

Have you made any contacts yet in the radio biz? (Better get with it, the book's almost over.) If you're working part-time at a radio station doing a little monkey work, ask the program director which school he thinks is the best. Remember, the guy in the trenches knows. After all, he's the one doing the hiring.

If you aren't working at a radio station, give a call to your local radio station and ask for the program director. He won't bite. You might not be able to get through, but it's worth a shot. Make contact with him, if you can.

Chug! chug! chug!

Some fraternities purport to be more than just drinking clubs. Not being a frat boy myself, I don't know for sure whether or not the top secret handshake will get you a job. If you decide to pursue a life drinking beer (and there is absolutely nothing wrong with that), here is a list of some broadcasting fraternities and organizations.

Contact the your local university to find local chapters.

- Society of Professional Journalists (Sigma Delta Chi)
- American Women in Radio and Television
- National Broadcasting Society (Alpha Epsilon Rho)
- Intercollegiate Broadcasting System
- Iota Beta Sigma
- National Association of College Broadcasters

Organization memberships get you more than a little newsletter and a bumper sticker that says, "Intercollegiate Broadcasters do it in 30 seconds or less." These groups are great places to network. You'll meet others in the field, including the up-and-comers. Then, it's just a matter of buddying up to them and riding their coattails all the way to radio success.

May I see your license please, sir....

For years and years and years, both radio announcers and engineers had to have FCC licenses. This, however, is not the case anymore. In fact, the FCC has made many changes recently. For example, the licenses used to be free, then they cost $45, then the goddamn things were free again, then they charged for them. Now licenses aren't required.

I have no idea what the FCC will do next week, but you can probably contact your local FCC field office and find out what they are doing this year. I suppose it depends on how many millions of dollars in fines Howard Stern has to pay in any given fiscal year.

If you don't need a license, cool. Announcers used to be required to have a "Restricted Radio-Telephone Operators Permit." To get that, you had to call your local FCC field office and request Form 753. There was no test and no fee.

Licenses for professional technicians are not quite as easy, I'm sorry to report. In addition to the General Radio-Telephone Operators License, you must also pass a test on Element 3 of the FCC examination and file FCC Form 756. Since you'll have to pass a test, you'll have to learn all the stuff before you take it. That means, if you didn't learn it in college or trade school, you better go get one of those "prepare for this test or that test" books and sit down and study.

Okay, I just stuck this stuff about licenses in here because I didn't know where the hell to put it. It's not something that most people in or going into radio really worry about. It's just something that you might or might not have to get done.

The bottom line about training

So does your training really matter? Yes, it really does. If you have pipes and lots of on-air moxie that will really knock program directors off their feet, just go to a trade school. The quicker you get the piece of paper that says you know what you're doing, the better. If you want to sit back and soak up all the liberal arts stuff, pick your school carefully. If you can land a job without any degree at all, great. But remember, when you're a young thing and just starting out, nothing (besides experience) looks better on a resume than a four-year degree.

The quick career change

For those of you considering a career change, already having spent plenty of time in the trenches of the almighty work force, a trade school is a good place to get quick and easy radio smarts. If you already have a four-year degree under your belt, even better. Coupled with your work experience, a shake-and-bake diploma and a good demo tape will take you far.

However, please think back to the chapter where I told you about going in the back door of the radio station. If you already have a four-year degree and can get help from a kindly disc jockey or program director, then don't worry about more school. You would just be wasting your time and money. Just get in and then maneuver yourself into your favorite spot.

You'd be surprised at how many people move out of boring or even despicable jobs (lawyers) and into the fun world of radio. Sure, it's mostly kids starting out and that might be a little depressing. But the voice of experience counts.

If you've been putting off your dream of landing a radio job because you were wasting time earning a living or having babies, stop right now! Time to start living your dream, even if it means handing out buttons on a hot day at the fair or getting coffee for Mister Obnoxious on the Morning Drive. The goal is to be there, right?

Whatever it takes.

College Radio

Everything you ever wanted to know about college radio but were too busy smoking cigarettes and planning Earth Day rallies to ask

There is another great thing about college that can really help your future in radio. It's college radio. Although associated with Birkenstock-wearing, greasy-haired, nose-pierced freaks, this one-stop-shop can help you get on the air at a paying station.

When I think about college radio stations, I think about hour after hour after hour of crappy, pretentious musicians singing about how lousy the world is. "Oh, Jesus never loved his mother, the leaves are dying, I was abused as a child, the flowers don't bloom anymore, blah, blah, blah, whine, whine, whine."

Then, I think about the disc jockeys and announcers I have heard who are so full of themselves, decrying the horrible inequities in the world that are perpetrated by the ruling class. Then they complain about how these inequities trickle down to their college campus because the dean's office will not officially recognize the Students For A Free Tibet Society, or some other trendy crap. Cry me a river.

Listen to the way the management at one college radio station describes itself: "Keeping up with the changes in musical styles, we are a progressive alternative station. Having a progressive format, we simply alter our format to meet the demands of our listening community, providing an alternative to the mainstream, MTV sounds we're so used to." Could they be more full of themselves? Or full of shit for that matter?

Join up

With my jaundiced, biased view of college radio stations and how stupid they are on the table for all to consider, I must deliver this distressing advice to you. You should join them. I know that they're ostentatious pricks. Hey, they make me sick, too.

You, however, have to keep your eye on the ball. The truth is, working at a college radio station provides a nice bullet for your resume and also gives you some good, solid experience. You'll see, first hand, how things work at a radio station. You will just have to be strong and grit your teeth while you play songs from among such stellar college radio staples as Placebo, Lilys, The Queers, Lullaby For The Working Class, Protein, and Nerf Herder.

You can regain your sanity on the way home. It's worth it. No, I'm not saying you should warp your brain and dye half your hair purple and the other half green. Chances are, if you keep quiet and do a good job, you'll get some quality education and experience in radio at the college station.

There is always something to learn

The head of the broadcasting department will have some influence over the station. The inner workings of the station will vary. These and other influences vary from school to school. Usually, the main goal, aside from rescuing the huddled masses from patriarchal oppression, is to teach students some valuable radio skills.

One department head at a Midwestern university says to become a disc jockey his broadcast students must attend the first station meeting of the year, sign up for an interview, and then be selected. Usually, they accept only half the applicants each semester, depending on the number of disc jockeys graduating. Then, the students must complete the training courses:

- 12 training shows of three hours each;
- two request shows of three hours each;
- two road shows, each one hour;
- two productions (a public service announcement and another, like a sign-on, drops, etc.);
- one production room exam to make sure they know the equipment;
- an oral exam on broadcasting equipment to show they know how to use it properly;
- a final exam on the FCC-type questions;
- inventory several shelves of CDs.

"This station has 40,000 CDs, so the slave labor is nice," says the department head. In addition, disc jockeys are required to attend the monthly station meetings and weekly trainee meetings. They're allowed only two absences. If they miss one of each, they're dropped. "After a semester of performing six hours of work for the station, most of the student-trainees come out with a pretty good appreciation and respect for the station," says the professor.

"There are all kinds of psychological theories explaining it (cognitive dissonance theory, etc.), but if you make somebody work for something, he or she will love it when they finally get it. Kind of like frat hazing, but on a nicer level. And each of the tasks and requirements has a valid purpose. I've heard of other college stations that have much more rigorous training programs than ours, and others that just show you how to run the board, and then tell you to go for it," says the professor.

Move through the ranks with caution

As you put in time at the radio station, you will find some opportunities for advancement that will look great on your resume. Depending on how well you perform (and how well the college radio higher-ups like you), you could be promoted to program director or promotions director.

Take advantage of every opportunity for advancement because any feather in your cap is great. Just don't let it swell your head. Though different gigs at the station make for nice little bullets on your resume, they don't necessarily mean a lot unless you have the voice or on-air personality to back them up.

Here's another reason not to fall in love with yourself: *It's just college*. While you are trying to conquer the college radio station, remember to keep working on your real-life skills. So what if you're the big fat king of a very small hill? "There is nothing I hate more than some know-it-all, snot-nosed kid who comes in here and thinks he's going to change this whole station around," says one particularly jaded radio veteran. "So you were the big man on campus? That doesn't mean shit out here. It means even less if you don't know your way around the [control] board. I don't care if you were the program director for three years in a row at your college station. If you don't know what you're doing, stay out of my way."

Whatever the level of expectations and requirements at your college, it's very important to stay focused on your ultimate goal of landing a radio job. Take advantage of your opportunity. Do not get sucked into any weird subcultures at your station where you become an advocate for this or a supporter of that. Don't get mixed up with the college freaks who hate America and think the only way to change it is by broadcasting their personal agendas. Stick with what you are being taught and are supposed to be learning.

There are a couple of organizations that cater especially to college radio stations. These associations are valuable for networking, and they're good information resources for both students and radio stations. Find out if your college station is a member of either one, if not both, of these associations and join them. Use them to your advantage by going to their seminars, reading their official publications, and getting whatever you can from them.

Intercollegiate Broadcasting System (IBS)
367 Windsor Highway
New Windsor, NY 12553-7900
(914) 565-0003, fax: (914) 565-7446
www.ibsradio.org

A nonprofit association of student-staffed radio stations at schools and colleges across the country. IBS focuses on management, programming, funding, recruiting, training, and other operational and creative needs for college radio stations. This one is really geared toward the stations themselves, not individual students.

National Association of College Broadcasters (NACB)
71 George St.
Providence, RI 02912-1824
(401) 863-2225, fax: (401) 863-2221
www.hofstra.edu/nacb/

NACB describes itself as "an accessible resource for advice and information, and a venue for exchanging ideas and innovative concepts." The association organizes regular conferences to improve the skills of college broadcasters, and also tries to be a link between college radio and the real world.

4. Say it Like a Pro

How to talk for a living

Some radio professionals might say their job is terribly easy. "All I have to do is go on the air and say what I have to, then I'm done." It might be easy for the veterans, but for the rookies, it takes time and practice to get an easy, conversational, informational style down.

The heart of everything you have to do, whether you're delivering the news or introducing a song, is communicate. To be a successful broadcaster, you need more than good pipes. You need to be able to communicate effectively with your audience. You want them to feel like they are on your wavelength. Don't talk down to them or talk at them. Don't think about talking to a million people. The best way to communicate is to announce as if you are talking to just one person. This is a great way to calm your nerves, and it also gets the desired effect, which is "personal" communication with an entire group.

Communication seems like the easy part of the job. However, it should not be taken for granted. Think about the best communication you have. It's between you and your friends, right? You know each other, and you speak in a relaxed, easy manner. That's exactly what you have to do on the air. Communicate in a relaxed, easy manner.

If you're not the type of person who says, "We gettin' ready to drop da funk all ova dis crib, yo!" then don't start doing it once you get on the air or on your demo tape. It's

109

your own personality that should come through. Listeners tune in between the songs to get a sense of who is behind the mike. They want to know you. They can also pick out a phony, so don't be one.

You are trying to create a sense of intimacy between yourself and the audience. The audience, you realize, is several thousand people, but you want to make each person feel as if you are talking to him or her, individually.

With that in mind, when broadcasting, don't use terms like "everyone," "all of you out there," "you guys," or any other terms indicating you are addressing a large group. Keep it intimate. Say "you," "our," "we," "I'm." Stuff like that. By the same token, don't get too comfortable. Say "the" not "tha" never say "uh-huh" or any other "lazy" lingo.

Speak as you do in normal, everyday conversation. Don't mumble. Don't yell. An easy way to keep a conversational style, is to use contractions. You normally don't say things like "We are going to start a non-stop music set at the top of the next hour." Instead, you'd say, "We'll start a non-stop music set at the top of the next hour." Keep it smooth, keep it real.

Larry King sez....

Big-shot radio and television interviewer Larry King gives this advice for communicators in his book, *How to Talk to Anyone, Anytime, Anywhere: The Secrets of Good Conversation*:
1. You don't have to be quotable. Not everything that comes out of your mouth has to be one for the demo tape, or something that must be recorded for time immemorial.

The point is communication—just get your point across and you'll do fine.

2. Attitude counts. No matter how nervous or worried you are, you must stay confident in your abilities. "I think one reason I've had a certain amount of success in broadcasting is that the audience can see I love what I am doing. You can't fake that. And if you try, you will fail," says King. Stay enthusiastic.

3. Remember to take turns. "Careful listening makes you a better talker," advises King. You will learn more and have a more successful interview, if you listen to what the guest has to say.

4. Broaden your horizons. Expand your world view. Don't just live in the issues that interest you. Seek out others who have something interesting to say and learn from them. This gives your show more interesting subject matter.

5. Keep it light. "One of my cardinal rules of conversation is never stay too serious for too long," says King. Guests, stories, and radio shows are more interesting to the audience if they are fun.

6. Be the genuine you. "You should be open and honest with your conversational partners, as you'd want them to be with you," says King. "Be willing to reveal what your background is and what your likes and dislikes are. That's part of the give-and-take of conversation, part of getting to know people."

Getting your show together

Your radio broadcasts will contain basic information. This is the stuff you hear every time an announcer speaks, but you hear it so often you don't pay attention to what, exactly, the announcer says each time he opens his yap. Here's what the radio pros include in each broadcast.

- the time;
- the weather;
- station call letters and their motto, such as "The Best Music of the 70s, 80s and 90s" or "Your Number One Groove Authority" (or whatever sludge they slap on T-shirts to sell the station);
- song title and artist;
- your on-air name.

What time is it?

Personally, I prefer to look at my watch, but a lot of people rely on the radio for the time, especially in the mornings or late afternoon. Basically, people want to know the time when they are thinking about getting to work or about leaving work. With that in mind, you should announce the time often.

Don't just say: "It's 4:20." Spice it up a little, vary your delivery. You don't want people to be aware you are delivering the time every time you open your mouth. Instead of saying, "It's 4:20," try saying, "It's 20 minutes past four," "We're coming up on the bottom half of the hour." You get the idea. Vary your delivery, but don't forget you have to communicate. Your listeners have to understand what your clever delivery of the time means.

It's how friggin' cold out there?

This is a biggie. Listeners always want to know about the weather. Will it snow? Is it raining right now? What's the temperature? What about Hurricane Ashley? Will she unleash her yuppie-named wrath on us? You get the idea.

When you give the forecast, you don't have to give the entire five-day forecast, the temperature in Sarajevo, or a bunch of other shit each time. Give a short weather report with the time. "It's noon...50 degrees, with cloudy skies." Of course, you can personalize it a little more. You could say, "Bring an umbrella, those clouds out there could mean rain…your WTCB time is noon and it's 50 degrees."

You should vary your delivery, as you do with the time, and be as creative and personable as you can. "Snow fall continuing through Monday. It's 20 degrees right now with a high near 30." "Looks like a great day to call in sick. It's 75 degrees now and we're under clear skies. By noon we should be looking at mid 80s and continued clear skies."

Be creative, but also get the information across clearly. Don't be so clever that your listeners have no idea whether they should wear shorts or a snow suit.

Call letters

Besides the fact that it's an FCC requirement, radio stations like having their call letters repeated to the point of listener nausea. It increases recognition, and helps the ratings game. Also, you'll have to add that stupid station motto during every break. Personally, I hate those dumb-ass mottoes, but considering that it is a requirement of the job, it's also a good thing to put on your demo tape.

Artist and song title

I'm not going to insult your intelligence. You know exactly what this is and what it means. When you go on the air, don't simply say the artist and song title. Instead, vary it a little. Get creative, but be clear.

What should I call myself?

As you listen to the radio, you might hear WWYY announcer Steve Austin giving the time and temperature, and leading into a "non-stop music marathon." Well, the guy's real name is probably *not* Steve Austin. He's using an on-air name. (By the way, Steve Austin was the real name of the bionic man.) Many, but not all, radio professionals use on-air names. They want to keep weirdoes from stalking them, or they want to project a certain image, or their real names don't "ring." Of course, some radio professionals simply use their real names.

This comes down to personal preference. A guy where I live uses the on-air name, "Ton E. Fly." I want to throw up every time I hear him say it. But that's just me. I hate pretentious bone heads.

Say your name is Doug Johnson. You might decide to change your on-air name to fit the atmosphere at the station.

Country station	Doug Durango
Rap station	Def Doug
Easy listening	Doug Dulcet
Oldies	Dazzling Doug
Alternative	Spunkmeister
Pop	Dookie Doug
Spanish	Douglas Del Fuego

Maybe I'm not the best guy to suggest on-air names. I think they get a little too cute sometimes. Personally, I'd stick with my real name, unless it was something totally screwed up like Richard Head, Lester Mulgaokar, or Igor Vasilevetskiy. If you do decide to change it, pick something that people can easily say, something that they'll remember, even if it's just a regular made-up name.

Put it all together

Now, put all these elements together. You want to get all five—time, weather, call letters and slogan, artist and song title, and your name—worked into your break routine. Here are a couple examples of how this can work. As always, you'll want to vary this and customize it to let your own, glowing personality shine through.

"Foo Fighters on WFTC...Where We Play Music On The Edge...Good morning, Spazz with you during the rush hour commute...it's 10 before nine...50 degrees and partly cloudy."

"Coming up on midnight here on WFTC...I'm Spazz and we'll return to our Smash Mouth super block here on WFTC Where We Play Music On The Edge...it's 50 degrees, partly cloudy...back to 'Walking On The Sun.'"

You get the idea. Work those five elements in together. Keep it fun, creative, and communicative.

Be prepared

Radio might sound totally off-the-cuff and unrehearsed, but don't be fooled. A common adage in the industry is "The best ad lib is a written ad lib."

Write down everything that you plan to say on the air. This will save you from two horrible fates of radio: dead air and "um." When you hear dead space or "ummmm" or "uhhh," you just caught a broadcaster who is not prepared. Writing everything down doesn't mean you are without spontaneity or liveliness. Quite the contrary, you use the cards of information as a safety net. If you forget the temperature or what song you are introducing next, a quick glance at the card can save you some embarrassment.

Not only will your brain fizzle out now and then, leading you to dead air, distractions happen in spite of that blinking "On Air" light. It's a sensitive subject at many radio stations, but the control room isn't such hallowed ground. Sales people, program directors, managers, and others are popping in and out of the room while you're trying to talk to thousands of people and sound relaxed.

Dead air on the radio is the about biggest no-no in the biz. It doesn't have to go on for more than five or ten seconds to be a problem. Even one second of dead air is bad. The best broadcasters do whatever they must to get on the mike right after a commercial or a song is finished, and they would probably crawl over their dead mothers to start a song or an ad immediately after their announcements.

How to Make a Break

Now that you have the basics on communications and announcing, try putting some of these little details together. This is great practice for your first job. As I mentioned earlier, radio professionals write down everything they plan to say. Write down five "ad libbed" breaks, including time, weather, station call letters and motto, artist and song title, and your on-air name. Then, using your tape recorder, practice reading them. Remember, stay communicative, personable, and informative. Keep re-recording until you get it just right.

Now, add some spice to those breaks. What you just did is the quickie, basic minimum. The best in the business add their own little goodies and extras. These little extras mark you as a professional.

When you listen to announcers on the air, you'll notice that they usually add little extras into their broadcasts. Along with the components you just wrote, try to work in some news, sports scores, public service announcements, or tidbits about the recording artists.

• Public service announcements. You've heard lots of these before. Insert something about "The Jackson County Humane Society," or free blood screenings, or "remember to test your fire detectors." You don't have to write a book about each one, just write a couple of quick, informative lines. Mull through the newspaper for some event and or make one up.

• News and sports. While you're checking out the public service announcements in the newspaper, take a look at the top news stories. Read through the story and pick out the highlights of each one. The first few paragraphs in each story have the pertinent details. Boil an article down into a two- or three-line summation. Sports are easy. Look at the score, then simply say, "The Vikings were defeated 42 to 27 in Green Bay." It's that easy.

• Artist tidbits. These are nice little bits of information about the artists or the song. Maybe you want to mention that particular U2 song was a Number One hit in 1987. Or maybe, after playing a Jewel song, you want to mention when her next tour is coming to town.

Now, take your tape of the original five elements, and work in some of these additional tidbits. Keep it short—less than a minute—and, as always, lively and communicative.

Keep practicing this. When you feel like you have some breaks which are particularly marvelous, hang on to the script you wrote. You're going to be using your best work on your demo tape.

Make Your Voice Sound Purdy

To be a broadcaster, you have to have a voice people like to hear. Unfortunately, some *schmucks* decide that they want to be in radio, but forget that they have to work on their voices and related skills.

Do you mean that I have to have a great big, rich, booming voice to get a radio job?

Do I have to sound like that guy who does all the voice-overs for the movie trailers?

Hell no. But you do have to develop and condition your voice so that you sound good. You must have a voice that is pleasant and easily understood. Does every broadcaster have to consciously go out and practice these skills? No way. But *you* should. Every little bit helps.

Your voice sucks, dude

Sometimes it is easier to explain how to do something by first talking about what not to do. This probably goes against some touchy-feely, P.C., "build through positivity, not negativity" theory, but I don't care.

• Don't talk through your nose. Let your voice come from your throat, not your nose. A nasal voice annoys the shit out of people.

• Don't breathe with your chest. Remember that cartoon where Bugs Bunny is telling Elmer Fudd to breathe with his diaphragm, then blows his head off with a bazooka? No one will do that to you, but breathing through your chest won't do the job correctly.

• Don't talk too fast. People are trying to listen to you and understand what you are saying. If you talk too fast, and they can't understand you, guess what? They won't want to listen to you! Most people talk too fast when they're nervous. Just talk to the microphone as if it's your buddy.

• Don't mumble. I don't know about you, but I hate it when someone mumbles at me. Speak up, man. You have to speak clearly and concisely to be in radio. Practice a few tongue twisters, or read something out loud, deliberately emphasizing and articulating each and ev-e-ry syl-la-ble.

• Don't use a wacky voice. Doing character work is fine, but don't use a fake voice all the time. First of all, it sounds totally unnatural and it's a turn-off. Plus, it will strain your voice. Speak normally and clearly.

• Don't yell. You will use a microphone and you don't have to scream into the thing. Think about disc jockeys or callers or guests you've heard who speak too loudly. It's really annoying. Control your volume. Speak in a normal, everyday voice.

Every breath you take

I know you are not an idiot, but you probably don't know how to breathe right. It's not your parents' fault, it's not the public schools' fault, it's not even the television's fault. Breathing for radio is a different ball game. It's more than merely trying to stay alive.

So what the hell is a diaphragm? Your diaphragm is the muscle that draws air into your lungs, then pushes it out. Let that muscle push out your words. If you breathe with your diaphragm, your voice will sound great, even if you have the world's runniest nose. The key is to expand your diaphragm (the muscle located just under your ribcage), not your chest, when you breathe.

Try to breathe in with your diaphragm, and then push the breath out by contracting your abdominal muscles. Do it a couple of times. It sounds like a lot of work just to breathe, but it does what you need it to do. Compare it to learning to drive a manual transmission. I ground the gears on my old man's pickup truck until they were totally stripped, but now driving a manual transmission is second nature for me. That's how good broadcasters regard breathing correctly.

Plus, breathing with your diaphragm does great things for your voice. It opens up your throat more, and makes the vocal chords resonate nicely. By routing your voice through your throat (and not your nose), you won't sound nasal. Not only will your voice be richer, but you will be able to speak longer between breaths. After all, you are taking in more air. Don't believe me? Try this: Breathe in through your chest, and hold it as long as you can. Now, do the same thing, but expand your diaphragm this time and hold your breath. See? You can hold more air in your lungs if you use your diaphragm.

Breathe, two, three, four...

Here are a few easy exercises you can do to make sure you are breathing and enunciating properly.

1. Make sure you control how much air you expel when you speak. Put a lit candle in front of your mouth, about five inches away. Read a magazine or a newspaper out loud and make sure you don't blow out the candle.

2. Stand in front of a mirror and watch your diaphragm expand as you inhale and exhale. While you do this, remember how your body feels when you breathe this way. Do this a shitload of times, so you know what you should feel like when you are on the air.

3. I wish I could teach my old girlfriend this one. Open your throat and keep it open. Take a really big breath and slowly exhale. See how long you can make the breath last. Next, do the same exercise, but hold a vocal tone. Practice this one, it helps your longevity.

4. Pick up a newspaper and read a story out loud. Make a note on the paper where you ran out of breath. Every time you practice, you should get farther and farther. Obviously, this isn't some race where you turn purple and mumble until you win. Use common sense here. Make sure you sound natural. Use the same newspaper every day so you can keep track of your progress.

Do these exercises regularly. Do exercise Number One for ten minutes each day and do exercise Number Two for five minutes. Do the others every few days, and whenever you feel like you need to get your chops built back up.

True test of tuneful tones

If you want to test your pronunciation mettle, give this little test a try. On-air personalities tongue-wrestle their way through the fictional "WFMT Announcer Audition," or similar challenges, to make sure they don't sound like idiots when they get on air.

The WFMT Announcer Audition

The WFMT announcer's lot is not a happy one. In addition to uttering the sibilant, mellifluous cadences of such cacophonous sounds as Hans Schmidt-Isserstedt, Carl Schuricht, Nicanor Zabaleta, Hans Knappertsbusch and the Hammerklavier Sonata, he must thread his vocal way through the complications of *L'Orchestre de la Suisse Romande*, the *Concertgebouw Orchestra of Amsterdam*, the

Leipzig Gewandhaus Orchestra, and other such complicated nomenclature.

However, it must not be assumed that the ability to pronounce *L'Orchestre de la Societé des Concerts du Conservatoire de Paris* with fluidity and verve outweighs an ease, naturalness, and friendliness of delivery when at the omnipresent microphone. For example, when delivering a diatribe concerning Claudia Muzio, Beniamino Gigli, Hetty Plumacher, Giacinto Prandelli, Hilde Rvssel-Majdan and Lina Pagliughi, five out of six is good enough if the sixth one is mispronounced plausibly. Jessica Dragonette and Margaret Truman are taken for granted.

Poets, although not such a constant annoyance as polysyllabically named singers, creep in now and then. Of course, Dylan Thomas and W.B. Yeats are no great worry. Composers' names occur incessantly, and they range all the way from Albeniz, Alfven, and Auric through Wolf-Ferrari and Zeisl.

Let us reiterate that a warm, simple tone of voice is desirable, even when introducing the Bach Cantata *Ich hatte viel Bekümmernis*, or even Monteverdi's opera *L'Incoronazione di Poppea*. Such then, is the warp and woof of an announcer's existence *in diesen heil'gen hallen*.

It Ain't All Top 40 and Hot AC

It is very important, when you've landed that first radio job, that you're prepared to work a format that you don't like. Personally, I hate the ever-living shit out of country music. I hate every reverberating note from an autoharp, and I hate every fake-twanged verse about why some guy's wife left him, and why his dog, named Tickbucket, bit him on the way to the county jail.

Chances are, if you really, really hate a certain type of music, you will probably end up working with it at some point in your career.

In *Broadcast and Cable's* 1997 yearbook, 71 different formats are listed for the 18,136 radio stations sprinkled throughout the United States and Canada. Stations range from adult contemporary to Eskimo to news/talk to Tejano to Serbian. And if you are particularly fond of Portuguese radio, there are six stations you can choose from. Any reggae fans out there? Enough I guess for two stations to exist.

The most prevalent station format, I am sorry to report, is country music. With 15 percent of the radio stations in the United States broadcasting that crap, it's no wonder our society is so sick. Nevertheless, you must prepare yourself to work at a variety of different radio stations, and you have to be ready and willing to happily tolerate whatever kind of music you truly hate.

"You just can't limit yourself," says Ken Ross, a Twin Cities, Minnesota, disc jockey at an adult contemporary station. "Even though I love R and B music, I have to be able to handle whatever kind of format comes up."

Classical music listeners are snobs

I would have a hard time working at a classical station, because I'd want to tell each pretentious shithead off, in between Rachmaninoff's *Requiem For A Spirited Flautist* and Mahler's *Enchanted Music For Dead Children*. I think these demographics, which were reported in *Broadcasting and Cable* (August 18, 1997), give you a sense of how very different people are out there. Plus, if you do end up at a classical station, this will be your audience. Pay attention to these dicks and learn from them.

A study was conducted by a snooty firm that owns a bunch of snooty stations. Whom did they study? People who listen to classical music, and have the audacity to claim, "We're not snobs." Yeah, bullshit.

The study found that classical music radio listeners:

• prefer radio to any other medium. While they are 27 percent more likely than the average adult to listen to radio, they are 25% less likely to view cable, and 12 percent less likely to watch broadcast TV;

• are far more likely than average to enjoy sailing during their leisure time. They also are more likely to ski, and to play a musical instrument;

• skew "fairly" young; those who listen primarily to classical music are in the 35-44 age group (24%), followed by 45-54 (20%,) and 25-34 (16%);

• live mostly in households with incomes above $40,000; 32 percent live in households that earn $39,999 or less;

• read *Time*, *Newsweek*, and *Southern Living* more than average; their next-favorite magazine is *Money*. (Have you ever even heard of *Southern Living*, or even picked up a copy of *Money*?)

"They aren't snobs, though," claims the article. "While those in $100,000-plus households are 43 percent more likely to watch *Frasier* than the average adult, they also are 36 percent more likely to watch *Friends*."

Spare me.

Movin' on up

It's a fact that a disc jockey leads a nomadic life. One day he or she might work at a rap music station in the heart of Philadelphia, and the next day that DJ might land at some country music station in Phoenix. Or, if he is lucky, the Eskimo station in Alaska. In order to be successful, disc jockeys must be able to adapt quickly.

So, because we want you to succeed, because we love you, and because you spent the money to buy this book, feel free to take advantage of this phrase translator that is sure to help you, no matter where your career takes you.

Here is a new record from <singer's name here>

Country: Shee-yoot! Get ready to kick your heels up over this new one from Travis Judd.

Rap: Dang, boy! We be poppin' a fresh new groove from Ice Pick, yo! Peace, I'm out.

Classical: Settle in and ease back, this effervescent new arrangement of Rachmaninoff's No. 87 Concerto in B minor, conducted by Wilhelm Straussbauer of the Denver Philharmonic, proves to be a melodious cacophony of mellifluous intonation.

Oldies: Hey Cats and Kittens! We're spinning the platters that matter here, daddy-o! Let's take a trip up "Blueberry Hill" with Mr. Chubby Checker.

The time at the tone will be midnight....

<u>Country</u>: It's midnight, pardner, and now's the time to decide. If 'n you can't be with the one you love, love the one you're with. Either way, keep your radio tuned to KNTY.

<u>Rap</u>: Boy, I hope you know where your girl is, 'cause it's straight up midnight, G, here on WRAP.

<u>Classical</u>: The witching hour approaches with intonations of the hourly chime. When next you hear the melodious reverberation through your stereo, it will be midnight here at WSNT.

<u>Oldies</u>: Time to head for make-out point, hep cats. It's nearly midnight here on KBOM.

Especially when you are starting out, you must be willing and able to work at any radio station that wants to hire you. If you loathe classical music, but they're hiring, you might have to suck it up. At this point, you can't really be snooty and choose one station over another, just because you don't like the music they play.

Moreover, it's not just a matter of getting hired. You might be comfortably working at a station, then—wham!—the format changes drastically. You'll never see it coming. If you can't make the transition, you'll be out of a job.

Take some time and spin through the radio dial. (It's just jargon. I know the radio's got buttons and it's digital.) Stop on the stations you don't normally listen to and try to get a sense of what the disc jockeys are playing and saying. You never know when you'll find yourself working at the Eskimo station.

5. The Famous, the Not-So-Famous, and the Infamous

Whoa, I heard of him

The world of radio is full of success stories—it is even more full of unsuccessful stories, but let's focus on the people who made it, and made it big. There's an old adage that says an overnight success usually takes about 15 years. Please keep that gem in mind when you read about some of the bigwigs in this business. Like you, they started with a dream and the desire to get off their asses and chase it.

I wish I could tell you a sure-fire, this-is-how-you-can-hit-and-be-a-real-overnight-success story, but there really aren't any out there. There is no formula, no magic spell that will make you a megastar. The way to make it in this business is simple hard work and perseverance.

Big stars such as Larry King, Howard Stern, and Rush Limbaugh have been on top of the heap for a few years now, but they too are examples of people who worked long and hard. Not one of them started with a gazillion listeners, or a nationwide network of hundreds of radio stations, or wallets bursting with money. Everyone started small at local radio stations, getting paid shit, and doing his or her best to stay on the air and be successful. They all had their share of career breaks and big disappointments. They have all had great career days and shitty career days.

The only difference between the Sterns and Limbaughs and the rest of broadcasting America is that their particular hooks caught on at the right time. The next Stern, the next King, and the next Limbaugh are out there. It's a matter of luck, timing, and a jock who can develop his or her own unique style.

The Big Dogs

As a child, Limbaugh loved radio. He scored his first on-air job as a teen, and started a slow, steady climb to the top. Limbaugh was fired from five jobs before becoming a public relations mouthpiece for the Kansas City Royals, then he went back to disc jockeying. Today, an audience of over 20 million listeners soaks up his conservative, white middle-class diatribes on radio and (formerly) on the TV syndicated *The Rush Limbaugh Show*. Why do his fans (who call themselves "dittoheads") love him so much? According to Rush, it's because he is "the epitome of morality and virtue ...with talent on loan from God."

The dirty joke master

Howard Stern was horny to get on the air (and horny in general), so he geared his whole life toward broadcasting and radio. As a kid, he learned AM from FM from his radio engineer father, and pursued his lifelong dream during college.

He got on the air at Boston University, then landed a disc jockey job in Hartford, Connecticut. He went on to jobs in Detroit, and Washington, DC (where he first hooked up with sidekick Robin Quivers). But it was the New York airwaves that proved most hospitable to Stern's irreverent, kiss-my-ass-and-while-you're-at-it-why-not-show-me-yours style. Today, Stern commands an audience of millions of loyal listeners.

Let's consider the life and times of CBS-TV and radio personality, Charles Osgood. His radio program, "The Osgood File," is heard by over 12 million listeners on 393 stations nationwide. Osgood started his long relationship with broadcasting when he auditioned to be a news correspondent with ABC. "I was dreadful," said Osgood. "It's a miracle they hired me."

After four years at ABC, Osgood moved around (albeit in the big leagues of radio) from station to station, and ultimately wound up at CBS.

Five days a week, Osgood gets up at 2:30 a.m. and goes to the office before dawn to prepare his radio show, which he writes by himself. To cap off his week, Osgood uses Saturday to prepare for his television show which he broadcasts live on Sunday.

Osgood doesn't even have a degree or training in radio. He has a degree in economics. He says the best bit of advice he ever learned about broadcasting is to be yourself. "I realized I shouldn't try to be like anyone else," Osgood said. "The only thing I was any good at was being Charlie Osgood."

King Kong

Larry King barely graduated from high school (he was one point away from flunking out), and his first job was as a janitor of a Florida AM station. He eventually landed his own Miami morning talk show. At the same time, he did TV commentary for Dolphins football games and wrote entertainment columns in two newspapers.

In 1971, King was arrested for theft. He swiped some money he had been given to investigate the Kennedy assassination. Somehow he was able to claw his way back onto the air. In 1978, he began a midnight-to-five a.m. show on the Mutual radio network, which had 28 affiliates. Now he's heard on more than 400 stations.

Regular Joes

For all the Sterns, Kings, Limbaughs, and Osgoods out there, there are thousands of more broadcasters who do not boast listenerships into the millions. From average success stories to others a little more unusual, there is a lot to learn from the regular Joes who do this job without ever being the topic of the morning water-cooler conversations.

Follow that dream

Bill Worthington of WASH-FM in Washington, DC, describes how he landed his first job and launched his radio career. "I was in my third semester of college, hanging around the campus station with a whole lot of interest and not much knowledge.

"On a bulletin board in the hallway at the campus station, the news director for a Top 40 station in the market posted a note requesting interns for his news room. I removed the note, put it in my pocket and scheduled an interview. I auditioned some copy and was accepted for the position. I re-wrote news copy of the morning newscasts, and played around in production for about four months while my grades deteriorated. When the station's program director became dissatisfied with his late night jock, I was hired for the shift."

On air in America (via Japan)

Twin Cities, Minnesota, disc jockey Steve Sundberg of WLTE-FM took a route into radio more geographically circuitous than most. "I got my start in broadcasting when I was a high school student at an international school in Tokyo, Japan. A school chum moonlighted at NHK Radio, serving as a teacher's assistant on an English-language education radio program. She thought I would sound good on the radio, so when NHK was developing another similar program, she recommended me to the producers.

"I auditioned, got the job...and three weeks later was asked if I'd be interested in doing a TV program, too. I stayed with the radio show for three years, and was with the TV show, *A Step To English*, for a couple of years beyond my high school graduation.

"But it was through another American who moonlighted at NHK that I got my first disc jockey job. He had just left the Air Force where he'd been chief announcer at FEN (Far East Network). A local Japanese company had hired him to build an English-language cable radio operation. He needed somebody to work the overnight shift and, even though I lacked any jock experience, he thought I'd do okay."

Love the job

The success story of Mark Howell, News Director of KUZZ AM-FM/KCWR in Bakersfield, California, is based on a love of broadcasting. "I got my start 32 years ago by sheer persistence. Having known since about age 10 that all I ever wanted to do in life was be on the radio, I went to a small liberal arts college that had its own student-run FM station, but no broadcasting curriculum whatsoever.

"As soon as I got to campus I was knocking on its door. Soon I was on the air at WECI-FM, Richmond, Indiana, doing a soul-music show on Friday nights. The first night I was scared out of my mind, fumbling all over a homemade control board. But I got hooked! I'd never had so much fun in my life, and I wanted it to go on forever. My whole world became that radio station. Actual school work was just something I had to do to stay there.

"At about the same time, I went looking to make some money at one of the local small-market commercial stations. Fortunately for me, one of them was run by a hot-tempered alcoholic who regularly fired people for no reason, other than his having a bad hangover that day. I happened to hit him when he was desperate for a warm body to work weekends. So I then had a job as a screaming Top 40 jock on 500-watt WHON (AM), Centerville, Indiana. When I quit eight months later, I was the most senior member of the on-air staff. But no matter. I was learning my craft.

"By the end of freshman year, I was the manager of WECI, and had quit WHON to go to work at a pioneering and money-losing FM station, Rich Williams' WGLM, Richmond. The station was far ahead of its time, doing full-service personality radio on FM in 1966. Paychecks often bounced, and the place ran on a shoestring but, again, I was learning my craft.

"Four years after that first broadcast on WECI, I graduated from Earlham with a degree in political science. I hadn't had even one minute of classroom instruction in broadcasting. I had been station manager of WECI for two and a half years, program director of WGLM for a year, had covered major news stories, played almost every kind of music, and fallen totally, irrevocably, head-over-heels in love with radio."

Howell's experience echoes that of many other broadcasters. Persistence pays, along with a love, a real love of broadcasting. That is really what's necessary to get in the door and be successful.

Getting in early

I don't know if anyone should be impressed with or jealous of Eric Roberts (not the actor). As a broadcaster, Roberts had his start in 1994 in Hurst, Texas. There he interviewed the local mayor, council members, and sports figures. He started when he was 14. Roberts is now a field reporter of sorts for Radio AAHS, a network aimed at kids 12 years old and younger.

"It's kind of a rush," Roberts said in *Boys' Life* (January 1996). "If you have a good show, you get a pretty good feeling. When I first started, I was nervous before every show. Now, even though I still get a little nervous, we've gotten better at improvising and at coming up with funny things to say on the air."

No two people have made their name in broadcasting in exactly the same way. Let's hear what other broadcasting bigwigs have to say about success and achieving it.

• Ed Bradley of CBS' *60 Minutes* credits his success to hard work. "Being the first one in, and the last to leave."

• David Brinkley, retired news leviathan, says you can't communicate if you have your head up your ass (or words to that effect). "You can't communicate with others unless you understand what communication means."

• Deborah Norville (whose legs, I swear to God, go all the way up to the top of her head) says a balance between personal and professional life is extremely important. "Life is great, but don't take it too seriously," says Norville.

These success stories demonstrate the value of hard work and perseverance. Unlike Charles Osgood, your first radio gig might not be as an ABC correspondent. It might be small. It might be humble. But it's radio.

As if you didn't notice, all the people mentioned in this section are absolutely in love with radio. It consumes their minds and lives, and it's the only thing they want to do with themselves.

I'm not telling you to build a little radio station in your basement, perform make-believe radio broadcasts, and stop bathing altogether, but you do have to be absolutely obsessive about radio. It's a job requirement.

Ethics? Hey, I Got the Job

Whatever works for you

Not everybody in radio went to school for four years, scored an internship, and climbed the ladder. Some made it the old fashioned way—they got lucky.

As you read these stories, please bear in mind that I'm not endorsing anything that could be considered unethical. Acceptable standards of conduct should, nay, *must* be adhered to. The following stories are for entertainment purposes only. On the other hand, if you think it will help you land a job, go for it.

The race card

One unnamed job seeker, while in junior college, got it in his head that he wanted to work in radio.

"I had my mom take me to all the local radio stations. At the next to last one, I cut a tape while I was there. They told me that they weren't hiring, even though I said I loved radio so much that I would work for free. I was so disheartened that I didn't even go to the last one. Then the general manager of the station where I cut the tape called me in and offered me part-time work over the summer, which I took. That led to full-time work after graduation.

"The unknown piece of this puzzle is that I am a member of a minority, and the general manager was unable to tell that from my tape. He was looking for a "minority" who didn't sound like one."

Recognizing the racial implications of this, he adds, "And for the record, I got the full-time job because of how good I was, but I would *not* have turned down the part-time job had I known the impetus behind it."

Delivering more than time, traffic, and weather

Who says nothing good ever came from a part-time job? Kelly Lockhart, a Tennessee disc jockey, says, "My first radio gig, in morning drive no less, was at a 2500 watt FM adult contemporary station in Key Largo, Florida.

"How did I get the job? By being the regular Domino's Pizza delivery guy to the general manager. I got to know the guy, and even went to his house-warming party when he bought a new place. One day, out of the blue, he called me at work and told me that he'd fired his morning guy and wondered if I wanted the job. Eleven years later, I'm still in the biz."

Thanks, Mom and Dad

Nepotism isn't necessarily a bad word, especially if it helps you get the job you want. Consider the case of this radio pioneer who shall remain nameless. "I haunted every station in town, starting at age 13. When I was 17, my best friends knew someone who had just sacked his music director. My friends begged him to give me the job. Boy, was I shocked. He hired me! That was my start, and I'll always remember how grateful I was to him, and to my best friends...mother and daddy."

Do your homework

If you really want something, you'll pull out all the stops to make it happen. It might involve becoming a pest and bugging the program director until he finally gives in. Or it might involve getting to know him in a very personal way. No, no. I am not advocating sleeping your way into a job, although it happens all the time. However, one broadcaster from Milwaukee, Wisconsin, says knowing a little detail about the program director helped him land a job.

"I was going for a job at WBCS when I found out that the program director, who is now programming in San Francisco, was a huge fan of the comic strip, *Incredible Hulk*. So I came in with three things in hand: a tape, a resume, and the latest *Incredible Hulk* comic book.

"Got the job the next day."

Winning more than Supertramp tickets

A Kentucky disc jockey shares her unusual way into the business. She happened to win an on-air contest and went to the station to collect her prize. When she picked up the award, she also left a resume. She landed an entry-level job as a researcher for the station. From there, she moved up the ladder to an on-air position.

Cable access gold

In 1973, another Kentucky broadcaster says he worked on a cable access show for some friends, providing super-groovy music. At the same time, the local radio station's evening show was run by one guy—coincidentally, the same guy who owned all the records.

The disc jockey quit and left town, leaving the station in a lurch. "The station got wind of the cable show and called to ask me if I wanted to bring my records out and be on the air. The engineer stayed with me until I got my provisional

FCC license and showed me how to run the board. They thought I sounded so bad that they gave me a music bed to play behind me when I talked. They said when you can barely hear yourself in the headphones, turn the background music up a little bit more. The LP's were all that counted."

Play both sides of the fence

Washington broadcaster Ed Johnson says he used a little white lie to land not one, but two jobs. "My first two jobs in radio, I got at the same time. I applied at an NPR affiliate at my college. At the same time, I applied at a Top 40 station in my hometown about 35 miles away for a weekend board op job. Each station required experience, so I used the other station on each application as current work experience. I was hired at both, thankfully, so that if they checked my resume, they would have found that I did indeed have a job at that station. They never checked.

"That same year, ABC sports did a nationally televised football game at our school and needed some temp help. The requirement was TV experience, so I said yes when asked at the interview. That was it! I was hired, and gained valuable knowledge and experience."

Names You Might Have Heard

Here's another route on-air hopefuls can take to almost guarantee a job on the air. A small handful of radio "professionals" out there have never had a bit of formal training; an ounce of vo-tech; or a dollop of preparation before landing on the air. Of course I'm referring to those folks who, thanks to popular culture, were propelled into afternoon talk shows and morning zoo jobs just because they did something freaky.

Test your knowledge

Match the radio personality (numbers) with his or her claim to shame (letters).

1. Tonya Harding
2. Daryl Gates
3. Marv Albert
4. Jessica Hahn
5. G. Gordon Liddy

6. Ollie North
7. Fred Goldman
8. Kato Kaelin
9. Danny Bonaduce
10. Denny Mclain

A. A grieving father who cried like a little sissy girl every time a camera focused on him over the death of his child at the hands of whitey killer O.J. Simpson.

B. Former sportscaster who dressed up in panties and a garter belt and forced women to give him head. If they refused, he bit them on the back. One woman escaped by snatching the rug off his head.

C. Iran-Contra figure and ex-Marine who sold weapons to the Iranians (of all people) to raise money for his own little private war. Then he lied to Congress about it.

D.   Watergate—the *original* White House Scandal—strong arm who lived in a sewer for a while, relying on rats for sustenance. (Incidentally, he also busted Dr. Timothy Leary once, while working as an assistant district attorney in New York.)

E.   Got her groove on with televangelist Jim Bakker back in the '80s. Helped topple the religious empire of Bakker, and his horror-movie wife Tammy Faye. The woman with the claim to shame found herself on the air in Phoenix, Arizona, reaping the rewards of her affair.

F.   Carrot-topped former member of the Partridge Family who got busted for punching out a transvestite. He is currently in Detroit, previously in Chicago.

G.   Former chief of Los Angeles Police Department, forced to retire in June 1992. It was the end of a stormy year, beginning with the videotaped beating of driver-while-intoxicated Rodney King and ending with the Los Angeles Riots. His catchy one-liners earned him a special place in the hearts of druggies and blacks, including: "Casual drug users should be taken out and shot."

H.   World's most famous house guest who not only used up his own 15 minutes of fame, but began leeching 15 minutes of fame from others. He landed a job doing afternoons in Los Angeles on KLSX.

I.   The only ice skater who ever looked slutty on the ice. In conjunction with her meathead ex-husband and using a couple of knuckle-scraping, mouth-breathing Neanderthal buddies, she had skater Nancy Kerrigan knee-capped. She was given a short-term stint by a Portland, Oregon, station.

J.   Former Detroit pitcher who came out of federal prison after a felony cocaine bust to eventually host mornings on Detroit news/talker WXYT. Eventually, he quit the station in a dispute with management, bought a thriving meatpacking plant, and is now the major employer in a small Michigan town west of Detroit. He looted the pension plan, bankrupted the company, put most of the town out of work, and wound up back in prison again in just a couple of years.

Answers: 1-I; 2-G; 3-B; 4-E; 5-D; 6-C; 7-A; 8-H; 9-F; 10-J.

The moral of the story is this. If you want to land a job on the air and don't necessarily want to spend all that time and money going to school, sending out resumes, making demo tapes, and getting rejection letters, all you really have to do is something atrocious that will catch the whole country's attention. Afterwards, just watch the job offers roll in.

The only one from the whole crowd who has a shred of credibility as a broadcaster is the most screwed up one of the bunch, Marv Albert. So the rest of them might not be the greatest on-air hosts ever, but who cares? They still got jobs. Of course, some of these "celebrities" have already spent their 15 minutes of radio fame and are now probably selling carpets.

Experts on Everything

Others who have made their way onto the air (without any radio training whatsoever) are experts in one field or another. Their background, skill, or whatever interesting hook they have, makes radio executives think they belong on the air. For better or worse, Dr. Laura Schlessinger and Dr. Dean Edell are but two of dozens of doctors on the air these days. I do think Dean Edell kicks ass, however.

Let's not forget Dr. Ruth Westheimer, the teensy, tiny sex therapist. Of course Dr. Ruth had her own *schtick*. First of all, she had that thick accent. I don't know if it gives her extra credibility or what, but people do listen to her. Second, she talked about sex!

It's a law of nature. If listeners think they're going to hear frank, graphic descriptions of sex, they will tune in. I know I do. Guys like Howard Stern, a true American who made our nation more lesbian conscious, wins zillions of listeners each day. One of Stern's biggest claims to fame and fortune comes from the sexual content of his show.

Now, apply this to someone like a Dr. Ruth. We're still talking about sex, but now it's legitimate. Because Dr. Ruth wants to help you get off, in a clinical, jargon-filled manner (she likes to say "erection" instead of "boner" and "vagina" instead of "honey-flavored love tunnel"), she's a relief from the fundamentalists and the others who hate talking about or hearing about sex.

Talk radio and call-in shows are popular because of these experts. After all, who is going to call in and ask some kid, fresh out of broadcasting school, what he should do about his no-account, cheating wife. (Unless the station is a country station; then the disc jockey would just play the latest song from Billy Ray Tucker or Zeke Terwillerger.)

Listeners rely on experts for some sort of reliability or competence. This can really suck for up-and-coming radio station employees, who get paid barely above minimum wage, and watch these "experts" get paid stacks of cash while they have spent time and money to go to school and now have to push the buttons for these prima donnas.

Journalistic integrity

Other experts who manage to get on the air without spending a second of training in front of a microphone are newspaper columnists. These writers have found fame and fortune in radio, especially in the wake of recent interest in talk radio and call-in shows. What qualifies them as experts in anything, besides meeting a deadline or writing about some governmental crap no one cares about? Beats me.

Let's take a look at a typical exchange between one of these great columnists and one of their mouth-breathing listeners.

The topic: Deferred tax credits for municipal bonding of non-regulated, incrementally duplicating fund restructuring.

The players: Dan Dobkins, columnist for the Witchita *Post-Intelligencer* and drive-time host of "Let's Chat" on KBBL, The Talk Station. And Irv Perkins, slaughterhouse worker and talk show fan.

Dobkins: Irv, you're up next on "Let's Chat."
Perkins: Dan? Is that you?
Dobkins: Yep, this is Dan Dobkins, you're on the air.

Perkins: Hey, great. I just wanted to tell you that I was right on with ya on yesterday's column about tax credits for the rich.

Dobkins: Thanks, Irv. What was your question?

Perkins: Well, actually, I just wanted to call and disagree with someone who called earlier. She said Roosevelt vetoed a law in 1940 that would allow public money to be used for nonprofit purposes. But really, it was Truman, many years later.

Dobkins: Hey, great point, Irv. That's good to know.

Perkins: Yeah, that sumbitch Truman didn't do a damn thing for the people. It really gets my goat when people call in and say such stupid things, ya know?

Dobkins: Yeah, I know, Irv. Hey, I'm up against the clock, I gotta go to a commercial break. Thanks for calling.

l don't know about you, but half of the crap they yap about on talk radio is so boring it makes my brain itch. But, talk radio has exploded as a news and information source. It has turned out to be a way in for many people, especially career changers who don't want to start at the bottom with the "kids."

No one spends a trillion dollars and nine-tenths of their brain going to medical school to start their own practice which, one day, will lead them to a talk radio gig. But, as I pointed out at the beginning of the book, decent people sometimes want to be something besides doctors.

Ditto newspaper columnists. No one sticks out 15 or 20 years in some low-paying, low-level newspaper job with the hopes that one day their dream of being on the radio will be realized. It just happens to lead them there when they're sick of what they're doing.

So how does the issue of radio non-professionals taking your job affect you? I'm glad you asked.

Among countless other obstacles and hurdles, you must develop a thick skin against this sort of thing. Chances are, at some point in your career, you will end up either working with or just being around one of these "pros" who come in and jabber about taxes or investments and give you a big headache in the process. Then you get to watch him sprint to the door a second after the show is over so he can go cash his check.

You, on the other hand, will have to go into work early, organize the celebrity's work for the day, line up guests, do show prep, and have a cool glass of water on hand while he or she is on the air. Additionally, you get to screen calls from wackos who know who killed John F. Kennedy.

Don't forget it's your responsibility to keep those commercials lined up, and press the buttons for the host, because God knows she hasn't a clue what to do and no inclination to learn.

Okay, I have played up the negatives of working with experts a little too much. Sorry about that. The important thing to remember is you must develop a thick skin for yet another of this career's little inequities. Some people might get a thrill or two from working with a big, hot-shot quasi-celebrity.

This is another "do your absolute best" spot. Help make the expert bring your station to number one in that time slot, and you become a commodity for the radio station. The expert is using you, so you use him to advance your career. That's how it works. Pretty soon, Ms. Sex Expert or Mr. Financial Wizard's name starts to look real good on your resume.

6. The Radio Resume

You, on paper

There are 18 trillion books and on-line resources to help you prepare a resume, so I'm just going to touch on the high points to help you get started. If you need more help or want a deeper, more philosophical explanation about each category, I suggest going to your library, your college book store, or checking in at the placement office.

You must do a professional job on your resume. Even though the career you are going into has virtually nothing to do with writing and everything to do with how you sound, a good resume is essential. If it's well-prepared, well-written, well-typed, and well-presented, it will mark you as a professional and will convince the program director to listen to your demo tape.

Use a word processor (this *is* the Information Age for christsakes) and a laser printer. Don't use your old eight-pin dot matrix machine. Print your resume on white, bond paper. If you feel the burning need for colorful paper, use the most conservative color you can, such as cream or very light brown. Don't use bright green or purple paper because you think it will stand out. It will stand out, but it will also find its way right into the trash.

If you don't have a computer and desktop publishing equipment, you should find someone who has a good setup or join the rest of us in the modern world and buy a computer. In fact, it might not be a bad idea to use a resume writing service. They'll put it together, slap it on nice paper, and give you a few dozen copies of it.

Your resume is your first impression. If it sucks, your tape could go right into the circular file. Program directors are looking for professionals. They aren't going to read some shitty resume written in crayon that says "I am good radio guy. You hire me now." A shoddily prepared resume (or none at all) tells the employers that you are not a professional. It also tells them you are careless, and don't give a shit whether or not you get this job. So do yourself a great big favor and do a good job on your resume.

Within each resume are sections on education, work experience, and stuff like that. There are certain areas you should spruce up a little more for a radio career.

Broadcast experience

Use the term "broadcast experience" instead of "work experience." This helps create an image of you as a radio professional in the program director's mind. If you have any radio experience *whatsoever*, include it. If you are mentioning radio stations, include the stations' call letters, cities of license, and frequencies. Detail what you did at each job and how long you did it, but don't write a book about each one. Trace your jobs back at least five years. However, if there are other jobs that would help present you as the best person for this job, make sure you include them. Include any jobs where you showed managerial abilities, a good work ethic, or special skills.

• Put your entries in this order: the title of your position, name of the employer, location (city, state), and dates.

- You should describe your work responsibilities, with an emphasis on achievements. This tells the employer about your skills.
- Identify your relevant work experiences and describe them fully. Don't include any irrelevant work experiences or keep them very brief.

Education

Here's where you put all the information about what you know. Document your college or technical school. Also include any extra courses (such as the free how-to courses that they give at public access stations). Make sure you list any degrees or certificates you received when you finished the course. List any relevant seminars or conferences that you've attended. For instance, if you attended a seminar on commercial writing, that's good. If you attended a driving night school course because of a DUI, that's not good.

Personal information

Finally, give some personal information about yourself. List any special skills that you haven't already included. Mention charity and volunteer work you might have done (even if it was *forced* volunteer work to get your diploma. Might as well use it now). Note if you are a member of a fraternity or sorority. Let the program director know about any impressive organization memberships. In fact, why not send in your fees and join a few now?

Hobbies are always great to mention, and they reveal your personality traits. If the program director and you both like playing softball, golfing, and playing Frisbee, this can give you a little edge over the competition. Obviously, when the guy hires you, he's not going to say, "Since you are such an avid duffer, like myself, I think you'd be perfect for the job." But it could give you a little subliminal push.

Computer and electronics skills are incredibly important to include. After all, radio is a technical business, and this may be the one little thing that wins you an edge. Make sure you shoot right down the middle in this section. By that I mean, don't exaggerate your skills, or leave out anything that can help you. If you know how to run the latest Windows workstation software, be sure to say so. But don't stretch it and say you wrote part of the program.

Spelling counts

Spell everything in that sucker right! Nothing screams "unprofessional" (or buffoon) louder than a misspelled word. It shows a lack of attention to detail, and it could cost you the job. There are a couple of easy ways you can avoid these little stupid mistakes.

• Have a bunch of people read your resume. The more eyes that see it, the more likely the little mistakes will get caught.

• Read your resume backwards. Start at the bottom of the page and read from right to left. Since you are not used to reading this way and the words will not make sense, it is often easier to pick out errors.

• Look up any word you are even the least bit unsure of in the dictionary.

• Try not to repeat the same words and phrases again and again. Even though this not a spelling mistake, it's very annoying. Use a thesaurus.

Formatting

Lay out your resume so it is visually appealing and uniform. If you enter the information about one job you had, enter the information about other jobs the same way. Align the date columns, heading columns, and information columns. If you are using a word processor, don't go nuts using stylized fonts and pretty bits of clip art. Your best bet is to stick with a professional-looking font, like Courier, Times New Roman, or Arial. Stay away from italicized text, unless you are listing a show, program, or other title, such as a book title.

I concocted the following resume to give you an idea what your resume should look like. There are no hard and fast rules. Resumes should look good, contain the necessary information, and present you in the best light.

Gregory Pope
222 Bleecker St.
Walla Walla, WA 98224
(207) 555-4758

Objective: My career goal is to continue work as a broadcaster, with an emphasis on sportscasting. I hope to continue developing and improving my skills as a broadcaster.

Broadcast Experience:
- Promotions Assistant, WWWW-AM, Lawrence, WA, 1996-1997. Assisted the station's promotions department and organized and implemented various promotional programs. Attended local fairs and public events as a representative of the station.
- Program Director/Announcer, The University of Washington, 1993-1996. Produced and announced a popular

evening show for the university radio station. Served as the station's program director for my final semester.

- Intern, WWWW-AM, Lawrence, WA, 1993-1995. Helped producers research, organize, and record various commercials and public service announcements. Also, helped broadcast local high school football games.

Education:

- Bachelor of Arts, University of Washington. My major was broadcasting, my minor was computer science.
- Washington Federation of Broadcasters Production Workshop, certificate received.
- Internet Web Page Workshop, certificate received.

Personal Information/Skills:

- Computer Skills: Microsoft Office, Adobe PageMaker, CorelDRAW!, HTML, WAVtable Pro sound editor.
- Alpha Epsilon Rho member
- Association of College Broadcasters member
- Child of a Vietnam Era Veteran Scholarship recipient
- My personal web site is located at www.gregpope.com

References:

Michael Johnson, Program Director, WXYZ-AM, Lawrence, WA, (207)555-4567.
Lucy Mendez, Promotions Director, WXYZ-AM, Lawrence, WA (207)555-4568.
Sally Fetter, former employer, *Donut Shack*, Lawrence, WA (207)555-8734.

7. Cutting a Demo

Listen to this....

Though your cover letter and resume are important, your audition tape is crucial. Resume and cover letter get you through the door; the demo tape is what you want heard. It gives the program director an idea of your on-air talent. Based on this tape, he will either call you for an interview or relegate you to the circular file. (At least until this program director gets fired, then you apply again when there is new blood at the station.)

This might sound like a no-brainer, but it is very, very important to put the best of your best stuff on this tape. If something was a great bit, but a little gremlin got into the works, don't use it. It's competition time.

Most program directors will listen to your tape for no more than 30 seconds before they decide if they want to call you for an interview.

Format

Don't think too hard on this one. Just put your demo on a 10-minute cassette tape. You don't have to buy some super-duper Teflon-coated, indestructible, super-low-noise, ultra-high-frequency-modulation audio tape. Don't use a 90 minute tape either; the program director doesn't want to fast forward through miles and miles of tape to get to your commercial section. Don't think that you have to go out and track down a compact disc presser to make a stack of CDs to send to potential stations. Just use a plain old, every day cassette tape.

Make sure you label the cassette itself, in addition to the case. Tapes and cases get separated easily. If your name is on the tape, the program director won't have any problems finding your number when he wants to call you for the job. Also, make sure you label the cassette neatly. Don't stick your name on a little chunk of masking tape or some other crap. Use nice labels, and type your name on them. If you have a computer, design a great professional-looking label for the tape.

The tape should be about 10 minutes long, but don't worry if you go a little bit over or under. It should have two sections:

- Announcing (five minutes, the front half of the tape);
- Commercials (three or four of your best ads, the back half of the tape).

Content

Remember, put your best stuff on the tape first. The program director will make a decision as to whether he wants to interview you almost immediately. So always start with your "A" material.

Start your demo tape with a big bang. Get the program director's attention with the first bit. Make it something really clever and exceptionally creative. Bear in mind, however, that this can be a problem, if your really clever, creative bits are too off-the-wall or outlandish. In that case, it's best to start in a more reserved manner, focusing on good, solid professional material.

Come up with a clever bit that isn't too controversial or inane. Of course, "controversial" and "inane" are relative terms, and the program director may be freakier than you are, but it's best to play it safe.

Check out Appendix E to find show prep material on the Internet. The web sites offer gobs of information about music and news. Plus, you will find some funny bits that you can work into your demo tape.

When you decide to actually make the recording, you must do a professional job. Remember, you will be competing against people who already work in radio and have access to high tech equipment that can make clean, clear tapes easily. You don't want a tape that sounds like you taped it on your boom box in your bathroom with the toilet running. To that end, you need a production studio. You can find places to rent for $100. Or, if you are in school, find out if the school has any recording facilities you can use. If you do end up renting time at a studio, remember that time is money. Practice your ass off before you go in there.

Fill 'Er Up!

I'll give you a sample script for the announcing portion of your demo. I'm just throwing a few songs out here, but you can customize the play list however you want. You should make different demos for country stations, top 40 stations, adult contemporary stations, and any others you want to approach.

My fictional station, WXYZ-FM 99, only plays the big hits from the 1980s.

Play: the last couple seconds of Prince's "Purple Rain."

Say: "WXYZ FM 99. Prince hit number one with that back in 1984. "Purple Rain." It's 25 degrees with snow likely throughout the evening hours into the night. The Animal Humane Society of Johnson County reminds you to "share your love and adopt a new best friend."

Play: the last couple seconds of U2's "With or Without You."
Say: "All the best of the '80s. U2 with their 1987 number one hit 'With or Without You' on WXYZ."

Play: the beginning of The Police's "Every Breath You Take."
Say: "Here's Sting and his fellow Policemen with 'Every Breath You Take,' on FM 99 WXYZ."

Play: the last couple seconds of Duran Duran's "Rio."
Say: "Duran Duran on WXYZ. Good afternoon, I'm David Johnson. It's half past three on an absolutely gorgeous Wednesday afternoon. Ninety-five and sunny outside right now, and looks like we can expect much of the same over the next couple of days. The Springfield Mall is performing free cholesterol tests throughout the rest of the week...visit the mall or call 555-1234 for more information."

Play: the first couple seconds of Wang Chung's "Dance Hall Days."
Say: "FM 99 WXYZ with Wang Chung's first chart topper. Coming up this hour we'll hear from Michael Jackson, Paul McCartney, and the Eurythmics. They'll be part of a dozen-in-a-row WXYZ Power Play."

Repeat this format and vary the breaks between songs until you have four or five minutes of samples for the program director to savor.

As far as the commercial section goes, read three or four of your best, well-written commercials. Don't worry about adding sound effects or any odd post-production work, unless you can pull it off and make it sound professional. If it just ends up sounding like some punk banging on a coffee can, don't bother. Stick with a professional, well-delivered commercial for Colon Blaster Bran Flakes or whatever else you want to advertise.

Demo tape Do's and Don'ts

- Keep it short, between five and seven minutes. Don't fill the tape with a bunch of material. Program directors don't have the time. Keep it short, loaded with your very best stuff.
- Keep the tunes to a minimum. The program director wants to hear your voice, not music. Use only a couple of seconds of music, just enough for the program director to hear your ability to lead into and come out of a song.
- No dead air on the tape! In the world of radio, dead air is like printing pictures of sick puppies in a children's book. Dead air marks you as an unprofessional *schnook*, and the program director will probably wipe his ass with your tape.

Quality control

Before you make your demo tape, make sure you have the script all written out. Practice it a few million times so it sounds good, then make a version on your boom box. Yes, I realize earlier in this chapter I told you specifically not to do this, but at this point we're doing a rough draft.

If you have friends or family in radio, make sure they listen to your tape and give you their honest, painful, ego-molesting feedback. Take the comments you get seriously. They might have a perspective on the tape that could help you.

Turn on the radio and listen to a well-known announcer. Does your voice on the demo tape have the professional sound and the conversational feel that the pros do? If not, work on your tape some more.

Once you've ironed out all the wrinkles, go into the studio with your revised script, and put everything together.

Bravo! Now you have a great demo tape.

8. Writing a Kick-Ass Ad

I'm a capitalist and that's okay

A necessary part of radio work—indeed a necessary part of your demo tape—involves writing ads and doing a little production. This sort of work is not everyone's cup of tea. In fact, many broadcasters absolutely hate writing ad copy and doing production work, saying, "It is not my job." Others recognize it as part of their job, and something they must do.

If you are someone who doesn't want to be whored by big business or industry, if you have such powerful morals and scruples that you couldn't *possibly* bring yourself to write an advertisement, then you have a tough road ahead. If, on the other hand, you have no problem or compunctions about writing ad copy, read on. I'll tell you how to do it the right way.

The nut

You have to sell a product or service in 30 seconds, maybe a minute. Not only do you have to present the product or service, but you have to keep the listener's attention and hold on to listeners so they don't change the channel while you are talking about Johnson's Lincoln-Mercury or Ev-R-Fresh Dairy Farms' new cottage cheese. By the same token, the commercial must also appeal to fickle, nit-picky clients.

The client's position is valid: The commercial must sell the product. The thing can be the most artistic, funny, clever ad ever written, but if it doesn't sell anything, forget it. The client couldn't care less.

In order to write and produce the best ad you possibly can, consider these points.

• The commercial must be compelling enough to maintain attention and interest for 30 or 60 seconds.

• Keep it simple. The spot goes by awfully fast, and you can't get bogged down in a bunch of crap. Get the message across.

• The commercial has to state the message loud and clear. The listener can't come out of the commercial saying, "What the hell was that all about?" At the end of the thing, the listener should be able to say, "Okay, sure. There is a sale at the Bagel Hut. Honey-nut cream cheese is 35 cents off this week."

• Present the product or service in a positive image. You have to play up the big, salable points. For instance, if you are producing an ad for computers, you play up the speed and stellar customer service provided by the manufacturer. You don't mention, "Our hard drives are 15 percent less likely to fail over last year's models." Certainly, you don't lie or make up any false claims, but do whatever you can to pump up the image and necessity of the product or service.

• Tailor the message to the audience. Customize your ads to fit the station format. With talk radio, the audience has tuned in, specifically, to listen. The ad should be written so they are compelled to keep listening. With music radio, the you want music as a background sound. The ad should stand out from the background, and grab the audience's attention.

• Always put your selling idea first. No matter how zany and wacky and creative you are, make sure the product or service is the star of the ad.

• Make a commercial that will stick with the listeners after the ad is over. When they go to the store, the great, positive image of "Mr. Sugar Donuts" must compel them to buy that brand.

• Write for the ear and the eye. Give your ad an image that listeners can picture in their minds. Use sound effects and creative writing.

• Make sure that your sound effects are distinguishable. There's a thin line between the sounds of a brook babbling and a toilet running; between a loaf of hot crusty French bread breaking and a compound fracture. Make sure the listeners know what they are hearing.

• Funny doesn't necessarily mean effective. Humor is tough to write and can easily backfire on you. Try to sell the product first; then go for the laughs.

• Identify the product or service early. People are going to hear a lot of stuff in your ad so make sure that the product or service stands out, then repeat it at least three more times. Follow the advice of speech writers: tell 'em what you're gonna tell 'em; tell 'em; tell 'em what you told 'em.

• Bury legal copy in the middle. People tend to remember the last thing in a list they hear. If you tell them a bunch of legal shit at the end, that's what they'll take away.

• Have a memorable toll-free 800 number, preferably one that spells something. Putting it at the end helps customers remember it, and that's where the client wants it.

• Force a response. Tell the listener to get down to the car dealer or call this 800 number for a free brochure. Doing this places a need to do something in the listeners' minds.

Mr. Advertising

Maybe you thought you couldn't do it, but you can. Right now, all by yourself. Write three ads: 15, 30, and 60 seconds. Throw in some of these enticing elements:

- Big Ed's Watch Warehouse
- 25-percent-off sale
- This Wednesday
- Sale on everything in the store
- Fiftieth and Carlisle Avenue

Now check out the ads in a newspaper or a magazine and write 15, 30 and 60-second ads based on those. When you are done, take the best of what you have done, fine tune them, then use them in your demo tape.

Do's and Don'ts

- Pump up the positives of your client.
- Don't try to fit too much information in to the ad, only what you need to get the job done.
- Stay on time. The ads should be 15, 30 and 60 seconds long, not 16 seconds, not 28 seconds, not 63 seconds. Hit it on the button.
- Don't put too many sound effects or other "extras" into the ads. Except for jingles or unusually clever sound effects, you should spend your time pumping the product or service.
- Stay focused. Don't let yourself go off on tangents.

Announce the commercials in an easy, conversational style. Don't sound like you are merely reading a script. It has to sound like it's coming off the top of your head.

On a career-building note, if you have a great sense of humor or can do some stellar voices, commercials are a great place to showcase them. It's hard to work your funny voices into your regular announcing. Use them in your ads. See if your friends think they're funny, before you put them on your demo tape. In fact, get your friends and family to critique your whole tape before you send it anywhere. Remember, you're going to have to please a lot of people during your career, so you might as well start doing it now.

For more information

For more information on ad writing, visit the On-line School of Radio Broadcasting on the world wide web at www.knowledge.ab.ca.

9. The Cover Letter

Please allow me to introduce myself

Your cover letter provides you with a great opportunity to tell potential employers things about yourself that won't fit into the resume. In it, you can explain some of the reasons why you are so eager to be on the radio.

Are you drawn to radio because your Uncle Ed took you to work with him on weekends at the transmitter tower? Was it your mother's dying wish that you be on the air? A cover letter is the best place to introduce yourself, and tell the employer why you are the best person for the job. Even if your resume lacks something, an enthusiastic cover letter could get your demo tape moved up higher in the pile.

Naturally, you want to write your cover letter neatly and professionally. Personalize it as much as you can and make sure that you don't make any stupid grammatical errors. No one is going to hire you if you weren't paying attention in the third grade. Check it over carefully and keep it brief—no more than one page.

When you're personalizing your cover letter, there are three things that just don't work: force, arrogance, and humor. Force and arrogance piss people off, of course. As far as humor goes, employers might laugh and even pass your funny letter around at staff meetings, but even if they're all rolling on the floor, that usually won't get you interviewed. Save your funny bits for the demo tape.

Here are some tips to help you write your cover letter:

- Address someone in authority by name and title. When this information is impossible to uncover, use a functional title (Program Director, Assistant Station Manager, etc.).

- Tell how you became attracted to this particular radio station.

- Demonstrate that you have done your homework on the station and can see its point of view by discussing current problems, interests, or priorities.

- Convey your enthusiasm and commitment.

- Balance professionalism with personal warmth and friendliness. Avoid using generic phrases like "enclosed please find."

- Identify at least one unique thing about you, such as a special gift for getting along with all kinds of people.

- Stand out as an individual, but without coming across as gimmicky.

- Point directly to the next step, telling just what you will do to follow through.

- Remain as brief and focused as possible.

- Explain how you learned of the job opening, such as a networking contact or a newspaper ad.

- Be sure you understand the employer's needs, then state your training accordingly. Sell your skills as a qualified person.

- Describe your personality. Tell them how well you work with people, if that appears to be important to the employer. Use some specific examples whenever possible and try to fit your description of your personality to the requirements of the job.

- Indicate that you are enclosing a resume, transcripts, or other documents for the employer's consideration. The purpose is to show proof of the statements you have made about your qualifications.

• Request an interview. State that you are available for an interview at a time and place convenient to the employer. Do not forget to give your phone number.

Here's an example. Feel free to copy it, if you want.

Gregory Pope
222 Bleecker St.
Walla Walla, WA 98224
(207) 555-4758

Mr. David Jackson, Program Director
WTBC-FM Radio
1577 Main St.
Rhinelander, WI 54956

December 2, 1998

Dear Mr. Jackson:

Can you use a person who has
• a superb on-air voice?
• verbal skills, and the ability to deliver messages with impact?
• personal confidence, especially on the air?
• organizing and planning ability?
I have these talents, as illustrated in the enclosed resume. I would like to put them to work for WTBC-FM as an announcer.

Please accept this letter, my enclosed resume, and my demo tape as application for the overnight announcer position at WTBC-FM that you advertised in the Nov. 15th edition of the *Walla Walla Gazette*.

Last spring, I graduated from the University of Washington with a Bachelor's degree in Broadcasting, and I am confident I have the experience, knowledge, and enthusiasm required to meet your ongoing needs at WTBC.

My interest in radio began in junior high school when I began doing the morning announcements at Thomas Edison Junior High. In high school, I landed an internship with the local radio station, WWWW, where I did voice-overs for ads, and broadcast from the high school football games. These positive early experiences led me to pursue my college studies in broadcasting and to gain as much experience in the radio business as I could through various positions at the University of Washington station, WASH. In addition to my contributions as a disc jockey and announcer, I also filled in as a sound engineer.

I would greatly appreciate an opportunity to discuss my candidacy with you for the announcing position, and how I can become part of the WTBC team. I will call you next week to set up an appointment.

Respectfully,

Gregory Pope

Enclosures: Resume, Demo tape

Delivering the goods

Put your cover letter, resume, and demo tape neatly into a padded envelope. Make sure it's packed properly. If your resume shows up with the shattered remains of a demo tape, then you are screwed. Spell the program director's name right, and write his title correctly. Some of them like to be called "Operations Director" or maybe the particular fellow you're writing to is actually the "News Director."

Keep a good stock of current resumes, cover letters, and demo tapes on hand. As soon as you discover a good radio opening, immediately send out your package! Don't fart around. The post office can get your package to the radio station in anywhere from two days to a week, depending on how cheap you are. The more you spend on shipping, the quicker it will get there.

Try it your own way

Yeah, yeah, everything should be professional and neat looking; however, I can't resist sharing a couple of success stories of job applicants who took big risks that paid off.

I heard about one guy who put his demo tape, cover letter, and resume (which was neatly typed and well-prepared) into a pizza delivery box. Inside the pizza box, he put a smaller pizza box (with adequate postage) as a return mail envelope. When the program director received his mail that day, the pizza box was the first thing he opened. He laughed his way through the guy's demo tape, then called the applicant the same day and offered him the job.

Another one I love was about a guy who sent his demo tape, resume, and cover letter in a box full of really strong-smelling bubble gum. The program director opened the package, and listened to the tape while chewing a piece of the bubble gum. Even though he wasn't necessarily wowed by the applicant's tape, as more and more people walking by his office were drawn in for a piece of bubble gum, he got to thinking about the applicant. Later that day, he called the guy and offered him the job. That story blows my mind.

Check Out My Site

In the age of the Internet, many radio professionals are making their own web pages. Some people are very easily impressed with web pages, and a good one might help you get a job. To do this requires a little cyber savvy; however, there are programs out there that can help you get your web site up almost immediately. Or, you can take the long route, and learn Hypertext Markup Language. Me personally? I'd stick with the easy, cut-n-paste shareware programs.

Hot Dog Professional
www.sausage.com

This is nice, easy-to-use software that lets you slap together your web page, and post it through your local Internet service provider. The program is shareware which means you're supposed to send in a check when you fall in love with it. Hot Dog is the top dog in web page design tools.

Netscape Communicator
www.netscape.com

Netscape also includes an easy-to-use web page maker with the Netscape Communicator browser. This one is free, but it will eat space on your hard drive, so make sure you have plenty to spare.

10. <u>Find the Opening/</u> <u>Ace the Interview</u>

Just get me behind a live mike, okay?

This is probably the part of the book you've been waiting for. After all, what good is a bunch of ready information and materials, if you can't show them to anyone? How do those people in radio get jobs? How do they hear about openings? You never see ads in the paper that say "Ground Floor On-Air Opportunity." No one takes your hand and leads you to the station with the opening. And, as we all know, every program director's desk is swimming in resumes, tapes, and other assorted junk.

Don't be discouraged. It's not true that you can't get in unless you know someone. New people do get in. Here are some great places to find out about job openings in radio:

• Check the ratings for local radio stations. After you drool over the stations at the top of the list, start scanning down. The stations at the bottom of the list usually accept applicants with little or no experience.

• Apply at religious radio stations. Most of them are nondenominational, and probably won't care about your personal religious background. Plus, religious stations often hire people with little or no experience.

• Get an internship while you're in school. You can work at a radio station as an intern and learn the "behind the scenes" action. Sometimes (but don't count on it), your internship can turn into a full-time job.

• The classified sections in industry magazines are a great resource for job openings. Jobs will specify how much education or experience you need to apply, but forget about that. Instead, if the ad says you need a minimum of two years experience, send your package out to them anyway. The program director might think that your voice is perfect.

Tracking the elusive job

Keep track of your job hunting efforts in a separate notebook.

• On a sheet of ruled paper, make a chart listing companies, location, contact person, source of the job lead, date of contact, and comments. Add a space for follow-ups. This helps you keep track of where you sent demos and can also keep you from looking like a dolt if you forget to follow-up or if you happen to send two demos.

• Keep a journal of interviews and what you said. You might meet that person again for a follow-up interview or run into him or her at another station.

• Now that you've written all this stuff down, use it. Follow up on those leads, send out more resumes, call people back if you couldn't reach them the first time.

Don't think that keeping a notebook is a nerdy thing to do. It might be tedious but, believe me, you'll be glad you have it when you're flexing your brain, trying to remember if you sent your stuff to this station or that. Also, a job search notebook can help keep you pumped, as you check on it and work on it everyday. You'll be less likely to get discouraged, if you have one more call that you can make.

Stalking 101

For some people, landing their first radio job meant pestering the program director incessantly until he finally acquiesced. However, being a pain in the ass might not be as easy as it sounds. First, you could find yourself stuck when a secretary starts seriously screening your calls. Here are a few things you can try:

1. Be polite. Secretaries have to deal with jerks all day long. Don't be someone she *wants* to disconnect. You know how to be nice.

2. Get your act together. Make an outline of what you are going to say to the program director and rehearse it a few times. Make sure you sound professional. Also, don't waste the secretary's time by asking for information, such as the address or the program director's name that you can find somewhere else.

3. Call when the secretary's not there, during lunch or after she's gone home for the evening. Often, the program director will work later than the secretary. Unless you are kicked into some voice mail system, you might get lucky if the person who happens to answer the phone hands you off to the program director.

Nonchalantly, ask the secretary if you can have the program director's direct number. If she won't give it out, try another method. Most businesses have private branch exchanges, phone numbers set aside in blocks with the first three digits used to identify the company. For instance, if the station's main number is 555-5000, it's a safe bet the other phone numbers in the company are 555-5 (and then three other digits). Just start with 555-5001 and keep on dialing numbers until you get someone besides the program director's secretary. (Of course, you already know her number, so you're not going to try that one.) As soon as another person answers, you play the doof and say, "Oh, I'm sorry, I was trying to get in touch with Mike Johnson, the program director. What's his direct number?"

On the other hand, if the job listing says "no calls," then do not call. You don't want to shoot yourself in the foot by being overly eager. Also, don't tell the secretary that you are returning the program director's call if you're not. You'll just piss her off. You don't want her telling the program director that you're a big stinking jerk.

God bless the Internet

Cyberspace, here I come! You can pursue job postings throughout the world and even post your resume to different job banks so that employers can come knocking on your door. These sites can help you get your resume together or do a little job hunting. The Internet is a great resource, so if you're not already connected, get with the program, man.

Internet sites require different things of visitors. At most, you can simply search for a job. From there, you can mail in your resume and demo tape. Others might allow you to email your resume. Start at the broadcasting sites listed in Appendix C. From those, you will find links to even more sites that can open doors for you.

Ace the Interview

Speak up, son

Now that your resume is out there and your demo tape has been listened to and loved, the program director will call you for an interview. You have probably gone on job interviews before, but this isn't like landing a job at *The Piggly Wiggly*. You actually care about this job.

Here are some interview tips:

- Appearance is important. This is your first impression, a lasting one, so make it count. Dress well. Look spiffy.
- Have a positive, upbeat attitude. It will probably be the biggest influence on the hiring decision.
- Be on time for the interview, but don't arrive more than 10 minutes early.
- Fill out job applications neatly and completely.
- Be honest about all applicable areas of your life, experiences, and education.
- Be a salesman. Never exaggerate, but be confident in your abilities.
- Learn everything you can about the station and the company that owns it before you go to the interview.
- Good manners and common sense will always help. Plus, if you act like a dick, you aren't ever going to get hired for a damn thing.
- Ask for the job, flat out. Make sure they know you are really interested in working there.
- Maintain good eye contact.
- Present appropriate body language. Be relaxed and open, interested, and attentive. Sit up straight, with your feet on the floor.

- Concentrate on making your voice come across with vitality, enthusiasm, and confidence. Low tones convey confidence and competence; whereas high tones convey insecurity.
- Listen actively. Indicate that you heard and understood what the interviewer said.
- Choose your words carefully. The right words will come easily if you have done your homework.

Interview Don'ts

- Don't arrive more than 10 minutes early.
- Don't accept alcohol or cigarettes, if offered.
- Don't discuss controversial subjects.
- Don't promise miracles.
- Don't be pressured.
- Don't stress your need for a job.
- Don't interrupt the interviewer.
- Don't negotiate salary without a firm job offer.

As you get going in the interview, you will find that your love for radio will help you. If you're prepared, your strong desire to work in the biz will kick in and provide you with good answers and a great attitude. After all, here you are, right on the fucking doorstep! This is your big chance to walk through that door into a job that you know you're going to love.

Conclusion

WGIR concludes its broadcasting day....

The key to landing a radio job comes down to two major things: be persistent and stay optimistic. With those two feel-good nuggets stuck in your brain, use the tools in this book to find yourself a job in the radio business. Will you score a job overnight? Probably not. Radio is highly competitive, and for all the sucky misery that people seem willing to go through to become lawyers and doctors, most of them still want to do something fun. Also, it will take you some time to get your shit together and, ultimately, land on the air. But you can do it!

There are different ways to get into radio, different career paths, different job-seeking methods. Of course, you can combine those methods that I've mentioned to your advantage. For instance, I talked about scoring an entry-level, do-nothing job at a radio station sweeping the floors or sharpening pencils or something. This does not mean it is your only route in. Why not get a job like that while you're still in college or trade school? Then you'll have experience for your resume. Call it "production assistant" or something like that. Remember our "success story" broadcasters? Not only were they in school and working at the college radio station, but they picked up side work at local radio stations.

Your job search efforts and educational choices will have to come from you. You will decide which school is best for you; you will decide if you want to apply for a job halfway across the country.

Take a minute and think about why you want a radio job in the first place. Are you eager to hear your voice on the air? Do you want to write or sell ads? Do you love the funny bits you hear on morning shows? Dammit, those people sound like they're have such a great time *at work*. When you hit brick walls in your quest to get into a radio station, just remember why you wanted to do it in the first place. That's always a great way to keep your enthusiasm pumped up.

Pursuing a radio career is far more challenging than becoming a doctor or lawyer. For radio careers, it's not about working through some educational cookie-cutter. In radio, you must sell yourself, not a degree, to get the job. Half the lawyers in the country would drop their big legal briefcases in a minute, if they thought they could land a fun job in radio. Maybe some of them are reading this book....

So don't let the difficulty of the journey keep you from getting there. You've got the radio bug; it's in your guts. You know you want to do it, just as Howard and Rush and all the others knew at some point. Who knows who the big personalities are going to be in the next few years? You might be one of them. Or you might just get into a business that people love because going to work is like going to a great big fun party. All your friends are there. You're on the air. You've got your own gig. Man, for once, somebody's going to listen to you. This is the good stuff. Music, creativity, and excitement all around you.

All you have to do is...get into radio.

Appendix A: Publications

These trade publications can help you keep up on radio current events, and even score some job interviews. Most have classified ad sections. Many of these magazines are really expensive to subscribe to, so I'd recommend visiting them on-line, or checking them out from the library.

Advertising Age
www.adage.com
220 East 42nd St.
New York, NY 10017, (212)210-0100

Almost Radio Network
pages.prodigy.com/almostradio
On-line only. News, information, humor and show ideas.

Billboard
www.billboard.com
1515 Broadway, 15th Floor
New York, NY 10036, (212) 764-7300

Broadcasting and Cable
www.broadcastingcable.com
PO Box 6399
Torrance, CA 90504-9865, (800) 554-5729

CMJ Online
www.cmjmusic.com
11 Middle Neck Road, Suite 400
Great Neck, NY 11021

Contemporary Christian Music Magazine
www.ccmcom.com
107 Kenner Ave.
Nashville, TN 37205 , (615) 386-3011

Country Spotlight   On-line country music and radio magazine.
www.countryspotlight.com

Current
www.current.org
1612 K St. NW, Suite 704
Washington, D.C. 2006
A magazine all about public broadcasting. Job listings.

Earwig
www.earwig.com
On-line alternative music and radio magazine.

Gavin Report
www.gavin.com
Music research for radio.
140 Second St. , Second Floor
San Francisco, CA 94105, (415) 495-1990

GroovePlanet
www.grooveplanet.com
On-line urban music and radio magazine.

Guide to Talk Radio
www.talkradioguide.com
Post Office Box 342
Needham, MA 02192, (617) 433-0870
Lists syndicated talk programming.

iRADIO
www.iradio.com
PO Box 22875
San Diego, CA 92192, (619) 491 4876
Free by e-mail, $100 per year by regular mail.

Radioairplay
www.radioairplay.com
5217 Lansdowne
St. Louis, MO 63109-2308, (314) 481-4711

Radio and Production Magazine
www.rapmag.com
PO Box 170265
Irving, TX 75017-0265, (972)254-1100
Specifically written for radio production personnel and includes a
cassette each month filled with goodies.

Radio and Records
www.rronline.com
10100 Santa Monica Blvd. Fifth Floor
Los Angeles, CA 90067-4004, (310) 553-4330
The big cheese of radio magazines. To get at the good stuff, you must
subscribe. $300 per year!

Radio Personality Magazine
www.ftsbn.com/~rp
On-line magazine about the movers and shakers in radio.

Talkers Magazine
www.talkers.com
P.O. Box 60781
Longmeadow, MA 01106, (413) 567-3189

Appendix B: Organizations and Agencies

There is an organization, federation, sisterhood or fraternal organization for everybody under the sun.

American Federation of Television and Radio Artists
www.aftra.org
260 Madison Avenue
New York, NY 10016-2402, (212) 532-0800; and
5757 Wilshire Blvd., 9th Floor
Los Angeles, CA 90036-3689, (213) 634-8100
The web site features a jobs link. Also, scholarships are available to members or their children.

American Women in Radio and Television, www.awrt.org
1650 Tysons Blvd., Suite 200
McLean, VA 22102, (703) 506-3290

Association for Women in Communications, www.womcom.org
1244 Ritchie Highway, Suite 6
Arnold, MD 21012-1887, (410) 544-7442

Association of Independents in Radio, www.well.com/user.air
1718 M St. NW #361
Washington DC 20036, (605)341-0944, (888)YES-AIR7

Association of Internet and Radio, www.aironline.org
3535 Westheimer #226
Houston TX 77027, (888) 890-9AIR

Broadcast Education Association, www.beaweb.org
1771 N Street, N.W.
Washington, DC 20036-2891, (202) 429-5354
An organization for students and teachers.

Canadian Association of Broadcasters, www.cab-acr.ca
P.O. Box 627, Station "B"
Ottawa, Ontario, Canada, K1P 5S2, (613) 233-4035

Canadian Communications Foundation
www.rcc.ryerson.ca/schools/rta/ccf
350 Victoria St.,
Toronto, Ontario, Canada
M5B 2K3, (416) 979-5000 ext.7008

Canadian Disc Jockey Association, www.cdja.org
3148 Kingston Road, Suite 209
Scarborough, Ontario, M1M 1P4

Canadian Radio-TV/Telecommunications Commission, www.crtc.gc.ca
The Canuck's version of the FCC, CRTC
Ottawa, Ontario, K1A 0N2

Corporation for Public Broadcasting, www.cpb.org
901 E Street NW
Washington, DC 20004-2037, (202)879-9600

Country Radio Broadcasters, www.crb.org
819 18th Ave S.
Nashville, TN 37203, (615)327-4487

Federal Communications Commission, www.fcc.gov
1919 M Street N.W.,
Washington DC 20554, (202) 418-0200

International Webcasting Association, www.webcasters.org
Advocates of broadcasting over the Internet.

Museum of Television and Radio, www.mtr.org
25 West 52 Street
New York, NY 10019-6101; and
465 North Beverly Drive
Beverly Hills, CA 90210

National Association of Broadcasters, www.nab.org
1771 N St. NW
Washington, DC 20036, (202)429-5300

National Association of College Broadcasters
www.hofstra.edu/nacb
71 George St.
Providence, RI 02912-1824, (401) 863-2225

National Association of Farm Broadcasters, www.nafb.com
26 Exchange Street East
St. Paul, MN 55101, (612)224-0508

National Broadcasting Society, www.onu.edu/org/nbs
P.O. Box 1058
St. Charles, MO 63302-1058
A society dedicated to college student development in broadcasting;
also home of the Alpha Epsilon Rho honor society. Web site has a
career board and scholarship info.

National Federation of Community Broadcasters,
www.soundprint.org/~nfcb
Fort Mason Center, Building D
San Francisco, CA 94123, (415) 771-1160

National Telecommunications and Information Administration
www.ntia.doc.gov
c/o U.S. Department of Commerce
14th & Constitution, N.W.
Washington, D.C. 20230
White House adviser on communications issues.

Public Radio in Mid-America, www.prima.org
c/o KOSU-FM, 302 Paul Miller Building
Stillwater, OK 74078, (405) 744-6352

Radio Advertising Bureau, www.rab.com
304 Park Ave S.
New York, NY 10010, (212) 254-5472

Radio-Television News Directors Association, www.rtnda.org/rtnda
1000 Connecticut Avenue, Suite 615
Washington, DC 20036

Radio and Television News Directors Foundation
www.rtndf.org/rtndf
1000 Connecticut Avenue, Suite 615
Washington, DC 20036

Society of Broadcast Engineers, www.sbe.org
8445 Keystone Crossing, Ste. 140
Indianapolis, Indiana 46240, (317) 253-1640

Appendix C: Job Hunting on the Internet

Here are some great sites that everyone hunting for a radio job should visit at least once.

Airwaves Radio Journal Job Source Page
www.airwaves.com/job.html

Biz Radio
www.bizradio.com

Broadcast Careers
www.broadcastcareers.com

Broadcasters Training Network
www.learn-by-doing.com
This page can help you track down internships and mentors in the field.

BRS Radio Classifieds
www.brsradio.com/classifieds

Jockline Talent Listing
www.jockline.com

North Coast Talent Bank
nickanthony.com/jobs.htm

On-Air Job Tipsheet
onairtipsheet.com

Radio Design.com
www.radiodesign.com
This company produces custom resume packages for radio professionals.

Here are some other non-radio specific career sites that are also worth a visit:

America's Job Bank
www.ajb.dni.us/index.html

Yahoo: Employment Classifieds
classifieds.yahoo.com/employment.html

JOBTRAK
www.jobtrak.com

E.span
www.espan.com

On-line Career Center
www.occ.com

Career Mosaic
www.careermosaic.com

Career City
www.adamsonline.com

The Monster Board
www.monster.org

Career Resource Center
www.careers.org

College Grad Hunter
www.collegegrad.com

Appendix D: Usenet Newsgroups

Personally, I think most of these groups are filled with a bunch of in-love-with-themselves assholes who freak out if anyone dares to butt into their precious conversations. A reason good to do so! Or just read a few and find out what's going on in them.

alt.fan.art-bell, about Art Bell's weird late night talk show.

alt.fan.don-imus, about Don Imus.

alt.fan.don-n-mike, about Don & Mike.

alt.fan.g-gordon-liddy, about G. Gordon Liddy.

alt.fan.greaseman, about the Greaseman.

alt.fan.handelman, about Allan Handelman's Sunday-night talk show.

alt.fan.howard-stern, about the ultimate shock jock and King of All Radio.

alt.fan.jim-rome, about "The Sports Jungle."

alt.fan.joe-crummey, about the veteran Southern California talk host.

alt.fan.kroq, about Los Angeles station KROQ.

alt.fan.mark-brian, about Mark and Brian.

alt.fan.rush-limbaugh, about you know who

alt.fan.tom-leykis, about (duh) Tom Leykis.

alt.radio.college, about college radio.

alt.radio.digital, about digital radio.

alt.radio.networks.cbc, about the Canadian Broadcasting Corporation.

alt.radio.networks.npr, about National Public Radio.

alt.radio.pirate, about pirate radio.

alt.radio.talk, about talk radio in general.

alt.radio.whadya-know, about Michael Feldman's comedy quiz show on public radio.

alt.rush-limbaugh, more Rush chat.

alt.sports.radio, about sports programming on radio.

rec.arts.wobegon, about Garrison Keillor's *A Prairie Home Companion.*

rec.music.dementia, fans of Dr. Demento.

rec.radio.broadcasting, a great, general purpose newsgroup.

Appendix E: Show Prep Web Sites

There is an enormous number of resources on the Internet to help radio professionals prepare their shows. Visit any one of these sites to jazz up your demo tape or, after you land a job, to keep your show fresh and interesting. I make no claim as to the quality of these sites. Everyone's brand of humor is a little different. Honestly, the humor at some of these sites sucked, but everyone laughs at different stuff.

Almost Radio Network
pages.prodigy.com/almostradio
A cooperative show-prep service run by Cosmo, a morning host on New Jersey's WJRZ.

AM News Abuse
www.texas.net/~kaz
Funny news items

American Comedy Network
americancomedynetwork.com

Andy Waits' Radio Zone
www.ptw.com/~awaits
Show prep material and bios from the morning host on KTPI in Lancaster, California

Bad Dog Press
www.octane.com
Twin Cities, Minnesota, publishers who can provide real and "fake" guests for your show.

Big River Radio
www.bigriv.com
St. Louis firm that offers show prep for morning shows.

BitBoard Show Networks
www.bitboard.com
Show prep site run by Kidd Kraddick, the morning host at KISS FM in Dallas.

Bob Rivers' Twisted Tunes
www.twistedtunes.com/
"Zany" songs

Buddy King
www.cl-sys.com/buddy
Free show prep delivered via e-mail from the morning host at Oldies
94.3 (WLTA) in South Bend, Indiana.

The Complete Sheet
www.thecompletesheet.com
Prep from professional comedy writers and former big shot producers.

FACSNET
www.facsnet.org
On-line resource for journalists

Funny Firm
users.aol.com/funnyfirm/funny.htm
Comedy-prep service for radio personalities.

Guest Finder
www.guestfinder.com
Find the expert guest you need for any show.

Hickman Associates
www.flash.net/~comedy
Comedy and trivia show prep material.

HitsWorld
www.hitsworld.com
Hits World tracks the music charts for you.

Hot Topics
www.bookpromotions.com/hottopic.htm
A guide to the hottest talk-show topics, experts, and current books.

InterPrep
interprep.com

Show prep delivered via e-mail

Lifestyle Information
www.lifestyleinfo.com
Fax prep service that gleans its information from popular magazines and newspapers.

Milkman
www.worldlink.ca/~milkman
Even more show prep services.

Morning Punch
www.ccpunch.com
Show prep material delivered via e-mail.

PR Newswire
www.prnewswire.com
Get your fill of business news, via the public relations powerhouses at PR Newswire.

radioEARTH
www.radioearth.com
This is one of the best show prep and general radio information sites around. Site master Corey Deitz has an excellent array of show prep materials, radio software, plus the *Radio Rider* weekly newsletter.

Radio Lans
www.freenet.edmonton.ab.ca/~grady
More show prep.

RadioPrep
www.radioprep.com
Listings of available guests and paid programs from MediaPower, Inc.

Radio Space
www.radiospace.com
This site features news, features, interviews, and public service announcement ideas.

SINprep
members.aol.com/sinprep
Show prep, including "zany" news stories.

Sites With Audio Clips
www.geek-girl.com/audioclips.html
Need a sound from a movie, TV or cartoon? Here's the place to go.

Starlive
starlive.com
Hot websites and live Internet events.

Urban Artist Bios
members.aol.com/markndark/web/index.htm
Prep service for "urban" radio stations.

VPD2000 Global Prep
members.aol.com/BEdwa86493/vpd2000prep
/showprep.htm
British Columbia-based prep service.

WAV Central
www.wavcentral.com
Soundbites galore from TV, movies, and sound effects.

WAV Place
www.wavplace.com or www.wavplace2.com
If one site is busy, use the other. Two sites featuring, you guessed it,
audio clips from movies, commercials, blah, blah, blah....

White Stuff
www.infi.net/~cwhite
Comedy service

Bibliography

King, Larry and Bill Gilbert. *How to Talk to Anyone, Anytime, Anywhere: The Secrets of Good Conversation.* New York: Crown Publishers, 1995.

Zimmerman, Caroline. *How to Break Into the Media Professions.* Garden City, NY: Doubleday and Company, Inc. 1981.

Morgan, Bradley and Joseph M. Palmisano. *Radio and Television Career Directory: A Practical, One-Stop Guide to Getting a Job in Radio and Television.* Detroit: Visible Ink Press, 1993.

Ellis, Elmo I. *Opportunities in Broadcasting Careers.* VGM Career Horizons, 1992.

Connolly, Gary. "The On-line School of Radio Broad-casting," (20 November, 1997) www.knowledge.ab.ca.

Index

Acknowledgments

Unfortunately, no one ever reads the acknowledgment page of a book. I know this because I have only read one acknowledgment page in my life and that was so I could figure out how to write this one. Regardless, I have to acknowledge the help of several magnificent people who helped me get this thing written.

First, I thank all the radio professionals mentioned throughout the book who gave so freely of their time, expertise, and wisdom. I was not only impressed that they would take the time to talk to me, but they didn't ask for any money. Such is the sign of a true professional. Donna Hamel will always have my deepest gratitude for setting up the aforementioned interviews. It always helps to know someone who knows someone who knows someone. Also, I want to sincerely thank all the radio professionals who, via the Internet, shared their personal stories and opinions about the radio business.

My best friend, Eric Sorenson, helped me with the day-to-day ups, downs, ins and outs of authoring. Both Eric and my grandmother, Rowena, performed a ton of editing, so you didn't end up reading things like "that guy on the radio is real, real, ReaL, good. and stuff".

Finally, thanks to everyone who buys this book. You're welcome to send any praise, criticism, comments, or sick jokes to my e-mail address: babus@winternet.com.

About the Author

Once told by a high school creative writing teacher that he would never amount to a healthy shit as a writer, Robert C. Elsenpeter made it his mission in life to write at least one book that he could shove back in the wicked bitch's face and laugh heartily while she called him "Big Daddy Bobb."

Always a radio enthusiast, Elsenpeter dreamed of on-air life while growing up. Even though a change of heart led him to pursue writing, he now lives out his radio dream vicariously through his friends who work on the air.

Robert Elsenpeter is the winner of *The 1998 Society of Professional Journalists' Page One Award*. He works as a newspaper reporter and, in his free time, volunteers as a high-school speech coach, helping youngsters become better communicators.

He and his wife Janet live in Minnesota.